ENDORSEMENTS

" ...ng to happen in some distant future, in 2050 or 2100 or when computer projections tell us the glaciers may have melted. It is happening now, all around us. And it builds on and intensifies long histories of extraction, exploitation, extinction, and genocide. How do we fight despair, nihilism, and an eco-fascist politics of the armed lifeboat in the face of this gathering but unequally experienced storm? In this collection of urgent essays, the Out of the Woods collective argues that hope emerges from the acts of solidarity in the face of crisis that they term 'disaster communism.' Surveying four key terrain of social struggle around the ecological crisis—borders, nature, futures, and strategies—Out of the Woods plot an environmental politics grounded in antiracist, decolonial, and anticapitalist movements and solidarities. There is no better guide to building a future of collective possibility out of the ruins of the present than *Hope against Hope*."

> —Ashley Dawson, author of *People's Power: Reclaiming the Energy Commons* and *Extreme Cities* and *Extinction: A Radical History*

"This collection from Out of the Woods represents some of the most refreshing thinking on the politics of climate change and ecology. It is also the record of a collaborative project that arose from the dynamic conjunctures of radical theory and social movements, written for understanding and worldmaking rather than clicks or commerce. As climate change emerges as an undeniable fact demanding of solutions, *Hope Against Hope* is an important contribution to an honest examination of the interests, desires and futures which might be served by the range of answers offered across the political spectrum."

> —Angela Mitropoulos, author of *Contract and Contagion: From Biopolitics to Oikonomia*

"*Hope Against Hope* is an experimental book that shares important initiatives and dreams to work against the hell of current capitalist climate catastrophe and the worlds beyond that this hell will create. In doing so, it brings hope back from the future. It encourages and nourishes, through friendship and courage, a revolutionary present badly needed today. Oh, yes, Out of the Woods claims, a new world already exists and is set to abolish the present state of things!"

> —Gustavo Esteva, author of *Grassroots Post-Modernism: Remaking the Soil of Cultures* and *The Future of Development: A Radical Manifesto* and founder of the Universidad de la Tierra, Oaxaca

"Work, energy, and the planet's ecology are all in crisis—apocalyptically, so. Survival seems impossible, hopeless, in fact. And, and, and. In the struggles against enclosures and borders, wasted natures and foreclosed futures, there are possibilities for going beyond capitalism, the state, and ecological crisis. To the array of competing strategies circulating throughout contemporary climate movements, Out of the Woods offers *Hope Against Hope: Writings on Ecological Crisis* toward the most revolutionary, interminable prospects of a hopeful future born of struggle."

—Kevin Van Meter, author of *Guerrillas of Desire: Notes on Everyday Resistance and Organizing to Make a Revolution Possible* and coeditor of *Uses of a Whirlwind: Movement, Movements, and Contemporary Radical Currents in the United States*

"Out of the Woods collective frontally attacks three major limitations in the strategies and tactics of dominant strands in today's climate change movement: national paths to salvation that ultimately rely on and perpetuate oppressive border regimes and global inequalities; the widespread perception of the need to shrink rather than increase the forces of production; and, last but not least, undue optimism about so-called 'stranded assets.' *Hope Against Hope* is also a twenty-first century affirmation of Marx's highly enigmatic conviction that 'Mankind thus inevitably sets itself only such tasks as it is able to solve, since closer examination will always show that the problem itself arises only when the material conditions for its solution are already present or at least in the course of formation.' Importantly, its approach is a non-sectarian, original, and thought-provoking attempt at grappling with, anticipating, and preparing for struggles to come."

—Kolya Abramsky, editor of *Sparking a Worldwide Energy Revolution: Social Struggles in the Transition to a Post-petrol World*

HOPE AGAINST HOPE

WRITINGS ON ECOLOGICAL CRISIS

OUT OF THE WOODS
COLLECTIVE

Hope Against Hope: Writings on Ecological Crisis
Out of the Woods (editors)

ISBN: 978-1-942173-20-5
LCCN: 2020930522
10 9 8 7 6 5 4 3 2 1

Out of the Woods Collective, outofthewoodscollective@gmail.com

Common Notions   Common Notions
c/o Interference Archive   c/o Making Worlds Bookstore
314 7th St.   210 South 45th St.
Brooklyn, NY 11215   Philadelphia, PA 19104

www.commonnotions.org
info@commonnotions.org

Copyediting by Erika Biddle
Cover design by Josh MacPhee / Antumbra Design
Layout design and typesetting by Morgan Buck / Antumbra Design
www.antumbradesign.org

Printed in Canada

CONTENTS

III. FUTURES

IV. STRATEGIES

ACKNOWLEDGEMENTS

Out of the Woods Collective (OOTW) gratefully acknowledges the following publishers for granting permission to reprint edited versions of OOTW articles in this publication. All other essays in this volume were previously published on OOTW's blog at http://libcom.org/outofthewoods.

"On Climate/Borders/Survival/Care/Struggle," originally published in *BASE Magazine*, June 23, 2017, http://www.basepublication.org/?p=474.

"A Hostile Environment," originally published in *Society & Space*, November 22, 2017, https://societyandspace.org/2017/11/22/a-hostile-environment/.

"Organizing Nature in the Midst of Crisis," published as "Human Nature" in *The New Inquiry*, January 27, 2016, https://thenewinquiry.com/human-nature/.

"Cthulhu Plays No Role for Me," originally published in *Viewpoint Magazine*, May 8, 2017, https://www.viewpointmag.com/2017/05/08/cthulhu-plays-no-role-for-me/.

"Postcapitalist Ecology: A Comment on Inventing the Future," originally published at *The Disorder of Things*, November 4, 2015, https://thedisorderofthings.com/2015/11/04/postcapitalist-ecology-a-comment-on-inventing-the-future/.

"Blockadia and Capitalism: Naomi Klein vs. Naomi Klein," published as "Klein vs Klein" in *The New Inquiry*, January 7, 2015, https://thenewinquiry.com/klein-vs-klein/.

"Disaster Communism: The Uses of Disaster," published as "The Uses of Disaster" in *Commune Magazine*, October 22, 2018, https://communemag.com/the-uses-of-disaster/.

INTRODUCTION

In October 2018, the Intergovernmental Panel on Climate Change (IPCC) released its latest report. It is a work defined by exhaustive detail and three exact and exacting conclusions. The first is that the global target set in the Paris Agreement of 1.5°C of warming would have far greater impacts than were previously anticipated.[1] The second is that these impacts would still be vastly preferable to those incurred by 2°C of warming: a sea-level rise of nearly half a meter by 2100; a massive increase in the proportion of the population exposed to severe heat; a decrease in marine fisheries by three million tons; a sixteen-percent loss of plant species; and a ninety-nine percent decline in coral reefs.[2] Perhaps most striking, however, was the report's third and final conclusion: the window for containing these clearly catastrophic consequences is rapidly closing. If warming is to be limited to 1.5°C, just twelve years remain in which to undertake what the authors call an "unprecedented" transformation of society. As NASA scientist Kate Marvel notes, 2030 is not a deadline. Climate change is not "a cliff we fall off—it's a slope we slide down. We don't have twelve years to prevent climate change, we have no time. It's already

1. Stephen Leahy, "Climate Change Impacts Worse than Expected, Global Report Warns," *National Geographic*, October 7, 2018, https://www.nationalgeographic.com/environment/2018/10/ipcc-report-climate-change-impacts-forests-emissions/.

2. Kelly Levin, "8 Things You Need to Know About the IPCC 1.5˚C Report," *World Resources Institute*, October 7, 2018, https://www.wri.org/blog/2018/10/8-things-you-need-know-about-ipcc-15-c-report.

here. And even under a business-as-usual scenario, the world isn't going to end in exactly twelve years."[3]

The authors of the IPCC report intended it as a "clarion bell"—an intervention which would "mobilize people and dent the mood of complacency."[4] Yet the reception of the report was, for many, defined not by decisive determination but desperate dejection. As climate activist Mary Annaïse Heglar notes, "Lots of folks who had never thought about climate change, or who thought it lived on some distant horizon, are now coming to terms with its reality, here and now. They're terrified. And sad."[5] In an essay published just a few days after the report, she describes how she came to comprehend the scale of climate change and how it drove her to despair:

> I knew climate change was real. I knew it was dire. I had an inkling that it was not far away. But I didn't know just how bad it was. I didn't know how many innocent—and I mean innocent—people were already suffering hideously. I didn't know how many people had been marked as allowable casualties because they were born in the wrong places under the wrong circumstances.... Where other people saw bustling crowds of people, I saw death and destruction. Even as I walked on dry land, I saw floods.... I worried about how we would treat each other in the face of such calamity. I doubted it would be kind. (I still doubt that, actually.)

Heglar suggests that for many, the initial shock of destruction is not met with resolve but with grief. The realization that people, creatures, and entire ecosystems have died, are dying, and will continue to die does not immediately lead to determination but melancholia. As Heglar puts it, "We're mourning our futures ... some of us are mourning our todays, even our yesterdays." The quantification of destruction does not instantly inaugurate action to prevent it, especially when what is being destroyed is so all-encompassing. Indeed, the devestation is so total that its most spectacular forms—floods, storms, fires—are poor metaphors for the true depth of the damage. To really grapple with the scale of the destruction involves attending to the slower, less eye-catching processes: the pollution and erosion of the soil; the felling of forests that bind the ground together; the extinction of creatures that feed on and were fed by the Earth. "Climate change" is the dominant

3. Quoted in Andrew Freedman, "Climate Scientists Refute 12-Year Deadline to Curb Global Warming," *Axios*, January 22, 2019, https://www.axios.com/climate-change-scientists-comment-ocasio-cortez-12-year-deadline-c4ba1f99-bc76-42ac-8b93-e4eaa926938d.html.

4. Stabroek News, "Fighting Complacency towards Climate Change," *Stabroek News* (blog), October 20, 2018, https://www.stabroeknews.com/2018/opinion/editorial/10/20/fighting-complacency-towards-climate-change/.

5. Mary Annaïse Heglar, "When Climate Change Broke My Heart and Forced Me to Grow Up," *Medium* (blog), October 10, 2018, https://medium.com/@maryheglar/when-climate-change-broke-my-heart-and-forced-me-to-grow-up-dcffc8d763b8.

description of ecological destruction, but this is not simply a climate crisis, it is an *ecological crisis*. The catastrophe is not just emission counts but rather countless extractions, exhaustions, and extinctions.

When Out of the Woods started writing, it quickly became apparent to us that merely comprehending the breadth and depth of the ecological crisis could be a destructive thing in and of itself. Distress and despair arise from beginning to grasp the cascading scales through which the ruining of so many living and nonliving things is underway. Such responses are not misplaced, for the ruination of these things negatively impacts the possibilities for collective life they may have once held. The spectacular apocalyptic images of climate change in the received narratives, moreover, figure this equivalence as an inevitability: the breakdown of the climate is the breakdown of society. As the waves roll in on the cities, it is assumed, societies will break down and survivors will fight each other over whatever remains, while looking to the state and the military for salvation. In the classic "eco-catastrophe" film *The Day After Tomorrow* (2004), for example, survivors seeking shelter on the rooftops of skyscrapers experience hope only upon the redemptive return to the city of the US Army and its helicopters. In *The Road* (2009), only the repro-normative filial bond between a father and his son is made to matter amid a nightmare world of cannibalism and despair. Out of the Woods came together, in 2014, in order to reject such privatizing responses to the conjuncture and to collectively formulate alternatives without, however, disavowing despair.

"When people feel something is really urgent, or crisis-oriented," Kyle Powys Whyte argues, "they tend to forget about their relationships with others. In fact, most phases of colonialism are ones where the colonizing society is freaked out about a crisis."[6] The fear of social breakdown amidst calamity is a colonial terror. This can only be a fear founded on a forgetting of existing relations and a figuring of crisis as something that has not yet happened. As Heglar notes, "when I hear folks say—and I have heard it—that the environmental movement is the first in history to stare down an existential threat, I have to get off the train. . . . For four hundred years and counting, the United States itself has been an existential threat for Black people."[7] No disaster is experienced by a unified "We." Likewise, no disaster is

6. Quoted in Emilee Gilpin, "Urgency in Climate Change Advocacy is Backfiring, Says Citizen Potawatomi Nation Scientist," *National Observer*, February 15, 2019, https://www.nationalobserver.com/2019/02/15/features/urgency-climate-change-advocacy-backfiring-says-citizen-potawatomi-nation.
7. Mary Annaïse Heglar, "Climate Change Ain't the First Existential Threat," *Medium* (blog), February 18, 2019, https://medium.com/s/story/sorry-yall-but-climate-change-ain-t-the-first-existential-threat-b3c999267aa0. We would add Indigenous people here too. See Roxanne Dunbar-Ortiz, *An Indigenous Peoples' History of the United States* (Boston: Beacon Press, 2014).

defined by a sudden disappearance of kindness. These are myths of liberal political theory and its ideological extension through pop culture. It is ridiculous to imagine that solidarity and generosity emerge only through the conditional guarantee of a state-enforced social order.

The historical evidence seems to confirm our political suspicions. Rebecca Solnit shows that the differentiated destruction of disaster is frequently defined by "an emotion graver than happiness but deeply positive" that creates "disaster communities" founded on mutual aid and collective care. These communities are not run by the state nor are they defined by the sudden evaporation of race, class, and gender. Instead, existing collectives of those most affected by disaster are expanded and elaborated to build new socialities of solidarity—whether in Mexico City after the 1968 earthquake or in New Orleans after Hurricane Katrina in 2005. In these cases, when the state appears, it is not to help but to restore its own definition of order. It clamps down on 'looters' repurposing vital supplies to share, and 'squatters' seeking shelter in abandoned homes. After the 1906 San Francisco earthquake the US Army was sent in: they murdered between 50 and 500 survivors and disrupted the self-organized search, rescue, and firefighting efforts that had spread throughout the city. "The disaster provoked, as most do, a mixed reaction: generosity and solidarity among most of the citizens, and hostility from those who feared that public and sought to control it, in the belief that an unsubjugated citizenry was—in the words of [Brigadier General] Funston—'an unlicked mob.'"[8]

Out of the Woods faces such calamity by insisting that we must not forget our existing relations with others. Amongst the working class, the racialized, the gendered, the colonized, disaster is met with self-organization, solidarity, and care. These collectives share in common their struggle and survival despite and because of the ongoing disaster of capital, race, gender, and colonialism. Whatever happens, their circulation of kindness is undoubtable. There will, of course, be no such kindness forthcoming from the brutal nexus of raciality, capital, and coloniality, forces which Denise Ferriera da Silva argues are "deeply implicated in/as/with each other."[9] But what else should be expected from operations which presuppose violence? The ecological crisis is a product of centuries of this system, of innumerable extractions and exploitations, indescribable enslavements and extirpations.

8. Rebecca Solnit, *A Paradise Built in Hell: The Extraordinary Communities That Arise in Disaster* (New York: Penguin, 2010), 35.

9. Denise Ferreira da Silva, "Unpayable Debt: Reading Scenes of Value against the Arrow of Time," in *The Documenta 14 Reader*, eds. Quinn Latimer and Adam Szymczyk (München, London, New York: Prestel, 2017), 92.

No true salvation can come in the form of a US Army helicopter. It is as a result of these striations that we insist on the importance of understanding ecological crisis as incorporating and being reproduced by what Ruth Wilson Gilmore refers to as "the state-sanctioned and/or legal production and exploitation of group-differentiated vulnerabilities to premature death." In fighting against this premature death, racialized, poor, queer, disabled and feminized subjects ensure that the response to ecological crisis is also group-differentiated. This response is one of deepened kindness from kin (or 'kith'—to revive an old word referring to affinities grounded in place and action, rather than genealogy) and redoubled oppression from oppressors. It is in this deeper "kithness" that we find a solidarity which may yet change everything. If connections can be built across spatial and social differences, beyond their current fragmented form, they might yet begin to construct the provisional infrastructures of a new world amidst the ruins of the old. Out of the Woods chooses to recognize such disaster communities as a space of possibility for communism in the midst of disaster.[10]

To talk of disaster communism in these terms is to take the ecological crisis as a disaster. In the *Oxford English Dictionary*, these are the first two listed definitions of "disaster, *n.*":

1. An event or occurrence of a ruinous or very distressing nature; a calamity; *esp.* a sudden accident or natural catastrophe that causes great damage or loss of life.

2. The state or condition that results from a ruinous event; the occurrence of a sudden accident or catastrophe, or a series of such events; misfortune, calamity.[11]

The concept of "disaster" is useful because it can collapse the distinction between event and effect—between ruination and the resulting ruins. Part of what makes the planetary ecological crisis so difficult to comprehend is its complex temporality; the disaster is simultaneously happening, has happened, and will happen. This is also part of what makes recognizing the scale of catastrophe so psychologically devastating. It is hard enough to accept some cataclysm that is yet to come. It is even harder to reflect on the rapid ruining of the past, present, and future all

10. Ruth Wilson Gilmore, *Golden Gulag: Prisons, Surplus, Crisis, and Opposition in Globalizing California* (Berkeley: University of California Press, 2007), 246.
11. "Disaster, *n.*," *OED* Online, Oxford University Press, https://www.oed.com/view/Entry/53561. [Accessed June 2, 2019.]

at once. Talking of ecological disaster offers a way to enfold these different times of ruination.

This enfolding capacity is evident in Neil Smith's understanding of disaster as a composite of risk, result, and response. His writing also offers another shaping element in what it means to think of ecological crisis as disaster:

> It is generally accepted among environmental geographers that there is no such thing as a natural disaster. In every phase and aspect of a disaster—(physical) causes, vulnerability, preparedness, results and response, and reconstruction—the contours of disaster and the difference between who lives and who dies is to a greater or lesser extent a social calculus. Hurricane Katrina provides the most startling confirmation of that axiom.[12]

In line with our claim that ecological crisis is group-differentiated, Smith notes that "disasters don't simply flatten landscapes, washing them smooth," but instead "deepen and erode the ruts of social difference they encounter."[13] Clyde Woods, importantly, extends this rut further into the history of "plantation capitalism," noting that "activists in New Orleans were very insistent that there was not just a disaster and people were taking advantage of it, there was a disaster *before* Katrina."[14] While disaster collapses the distinction between the process of ruining and the ruins it creates, it simultaneously deepens the differentiation between who and what bears the brunt of disaster and who and what does not.

What we call "disaster communism" is an immediately ethical and eminently practical response to this differentiated disaster. Such a statement is anathema to the liberal devotees of establishment responses to catastrophe. To them, disaster communism could only be a perverse and antipragmatic faith in things that don't yet exist, or a dangerous romanticization of practices that have already failed. Such critiques indicate not only an obvious detachment from the lived reality of disaster communities, but also a determined ignorance of the inefficacy of supposed liberal "solutions." While liberal critics will claim disaster communism is based on promises not practices, they will also maintain a strange silence about the fact the IPCC's own solutions depend on as-yet-uninvented technofixes. This can only be the fantasy of that which claims to be nonideological, and therein sits the most pernicious ideology. Disaster communism already exists—indeed, some components

12. Neil Smith, "There's No Such Thing as a Natural Disaster," *Understanding Katrina: Perspectives from the Social Sciences*, June 11, 2006, https://items.ssrc.org/understanding-katrina/theres-no-such-thing-as-a-natural-disaster/.

13. Smith, "There's No Such Thing as a Natural Disaster."

14. Clyde Woods, *Development Drowned and Reborn: The Blues and Bourbon Restorations in Post-Katrina New Orleans*, eds. Laura Pulido and Jordan T. Camp (Athens, GA: University of Georgia Press, 2017), xxiv.

have existed for hundreds of years—but is criticized as a radical fantasy, while the as-yet-uninvented technologies of carbon sequestration and geo-engineering are taken as matters of scientific fact.

Not content with things that don't yet exist, leftist critics of disaster communism might supplement these technofixes with spatial fixes or displacements of the climate disaster. In this vision of the future, nature is not to be mourned but *managed*. In our view, the imaginations of assorted Keynesians, Green New Dealers, and accelerationists tend to be constrained by a romanticization of labor-saving technologies and automation. We do not want to be mistaken for defending work, yet what these architects of the future cannot admit is that automation does not save labor-time as much as displace it. The automation of production only changes the *form* and *composition* of labor and the *places* in which labor is performed. In terms of form and composition, automation merely reorganizes labor-time so that a greater proportion is devoted to the intellectual labor of innovation. Such knowledge is eventually embodied in machines, which of course have to be built as well. As Marx convincingly showed in the longest chapter of *Capital*, "It would be possible to write a whole history of the inventions made since 1830 for the sole purpose of providing capital with weapons against working-class revolt."[15] This shift is undertaken, whether consciously or unconsciously, in pursuit of crushing the power of workers. By reducing the portion of "living labor" enrolled in capital (thus reducing the number of laborers who can go on strike, blockade the factory or abduct their boss), the "dead labor" of machines (who can be relied upon not to blockade or sabotage anything) compose the value of a product or energy system.

Just as it changes the form and composition of labor, automation also changes the places of labor. All technology requires energy but energy itself is harnessed through work.[16] Even renewable energy systems require new or recycled raw materials such as rare-earth metals, lithium, and copper to be extracted. The cheapest way of doing this is through the appropriation of raw materials and exploitation of labor, most likely in the Global South. Under capitalism, mining is an inescapably violent and toxic practice. Separating raw materials from waste increasingly requires chemical and biological "work" (in addition to human and nonhuman labor) to recover harder-to-extract reserves. Costs can be driven down if environmental protections are circumvented outright through bribery or, more commonly,

15. Karl Marx, *Capital: A Critique of Political Economy*, Volume I, trans. Ben Fowkes (London: Penguin Classics, 1976), 563.
16. George Caffentzis, "The Work/Energy Crisis and the Apocalypse," *Midnight Oil: Work, Energy, War, 1973–1992*, eds. Midnight Notes Collective (Brooklyn, NY: Autonomedia, 1992), 215–72.

structural adjustment policies. Mined materials are circulated along hypersecuritized global supply chains. International maritime shipping is said to compose up to three percent of global carbon emissions. Yet such maritime transport is specifically excluded from the transnational Paris Agreement, demonstrating the absurdity of contemporary international climate politics (albeit the International Maritime Organization has recently tried to mandate for cleaner fuels). Globally transported raw materials must, of course, be coordinated with one another in order to be transformed and assembled in factories. The latter, more conventional, sites of exploitation ultimately rely upon directly productive labor as well as social infrastructures of cheap food, clean water and care, all of which maintain workers' bodies. Finally, products shipped to their points of consumption are used for increasingly short periods of time before being discarded into landfills or recirculated as e-waste for one last gasp of value extraction.

Automation has already failed the vast majority of the population of the planet. While it has undoubtedly benefitted white colonial capital, the effects of automation on racialized, colonized proletarians have always been disastrous. The form, composition, and places of human and nonhuman exploitation, capitalization, and appropriation continue to produce what Marx called "surplus populations" or the "industrial reserve army"—massive groups of unemployed laborers whose existence serves to keep the cost of labor down.[17]

It is worth considering a visceral example of the consequences of this system. Foxconn, the electronics giant infamous for its exploitation of workers in China, cut more than 400,000 jobs between 2012 and 2016 through the introduction of tens of thousands of robots.[18] By 2020, the company plans to fully automate thirty percent of its production. While Foxconn's jobs are rapidly disappearing, its ecological destructiveness persists. In 2013, Foxconn was accused of releasing vast quantities of heavy metals into tributaries that feed the Yangtze and Huangpu—the two rivers that supply most of Shanghai's water. Locals told reporters of high incidences of cancer. They had stopped eating cuttlefish from the rivers or vegetables from the fields for fear of the health consequences. The central government had

17. On the relation between the "organic composition of capital" and the production of "surplus populations," see the extraordinary analysis by Marx in *Capital*, Vol. 1, Chapter 25, "The General Law of Capitalist Accumulation." Marx was not particularly hopeful about the political potential of surplus populations. By contrast, following Frantz Fanon and others, we see those populations rendered surplus by capital to be indispensable to any revolutionary movement today.

18. Cissy Zhou, "Man vs Machine: China's Workforce, Starting to Feel the Strain from Threat of Robotic Automation," *South China Morning Post*, February 14, 2019, https://www.scmp.com/economy/china-economy/article/2185993/man-vs-machine-chinas-workforce-starting-feel-strain-threat.

already relocated one entire community away from the area, apparently because of its unnaturally high incidence of cancer.[19]

This is part of a much greater problem in China. In 2013, researchers estimated that "between 8 percent and 20 percent of China's arable land, some 25 to 60 million acres, may now be contaminated with heavy metals."[20] China's surplus populations thus face a double disaster: there are no jobs in the newly automated factories and they cannot return to live on the land because pollution has made agricultural subsistence nearly impossible. Automation is not a solution to the ecological crisis. It merely intensifies the vulnerability of the surplus populations it creates, making them ever more dependent on resources that capital has already ruined.

Faced with the impossibility of surviving on land where nothing can grow, amidst factories where no one can work, in housing where no one is safe, it is not surprising that surplus populations are forced into migration. Under global capitalism, it is impossible to escape the processes that produce local ruins. Racialized proletarians who move from one country to another may find work and thus survival but they will still be exposed to the differentiated disasters of the ecological crisis. After Hurricane Irma in 2017, for example, most reports focused on the damage to Florida's agriculture. The sugarcane harvest was destroyed as was much of the avocado crop. The fate of the 300,000 migrant workers who tended the farms was almost entirely ignored.[21] Many Latinx migrants chose not to go to the hurricane shelters, fearful that operators would report them to immigration enforcement. Some could afford the expense of a motel room in a safer area but others had no choice but to try to weather the storm. The hurricane completely destroyed many of the mobile homes these migrants were living in, worsening a housing crisis that was already dire. In the aftermath of the hurricane, migrant workers desperately needed new cheap accommodation, yet the destruction of the farms made it impossible to find work and thus to pay rent. Their experience is typical of the differentiated disasters the ecological crisis wreaks on surplus populations. In search of work and survival, migrants are forced to endure new vulnerabilities and more limitations on their mobility.

The global ecological crisis is a catastrophe of extraction, exhaustion, and extinction which exploits human and nonhuman things. As Che Gossett has

19. Paul Mozur, "China Scrutinizes 2 Apple Suppliers in Pollution Probe," *Wall Street Journal*, August 4, 2013, sec. Business, https://www.wsj.com/articles/SB10001424127887323420604578648002283373528.

20. Mozur, "China Scrutinizes 2 Apple Suppliers."

21. Georgina Gustin, "Florida's Migrant Farm Workers Struggle After Hurricane Damaged Homes, Crops," *InsideClimate News*, October 17, 2017, https://insideclimatenews.org/news/16102017/hurricanes-florida-agriculture-migrant-farm-workers-jobs-crop-loss.

argued, "The caging and mass killing of animal life, the caging and mass killing of Black life, and the racial capitalism that propels premature death are connected in a deadlock."[22] The extermination or carcerality of Black people and nonhumans are in a coterminous relationship structurally necessary for political domination. Race and coloniality operate as an endless destructiveness that constantly feeds on and into itself.

A potent example of this can be found in the colonial history of the island Nauru, where the violence of colonial resource extraction is reproduced in the present brutality against racialized migrants. Nauru was initially colonized by Germany in the late-nineteenth century before being transferred to joint administration by Australia, New Zealand, and Great Britain after World War I. Nauru's value was in its extraordinarily abundant deposits of phosphate—a crucial ingredient of agricultural fertilizers. From the early twentieth century on, the British Phosphate Company strip-mined the island during a frenzied resource boom. For a brief period, Nauru had the second-highest GDP of any nation in the world—only Saudi Arabia was richer.[23]

When the phosphate was finally exhausted in the late nineties, the country fell into a deep crisis: the central bank went broke, unemployment hit ninety percent, and the school system collapsed. After decades of strip-mining, the very foundations of the island were in ruins. Geographer Anja Kanngieser writes, "the mine area, called 'topside' by Nauruans, is like a moonscape. Huge limestone pinnacles reach skywards, punctuated by steep gullies into which, I was warned, people have fallen to their deaths. It is unbearably hot, humid, and inhospitable."[24] Eighty percent of Nauru's surface was now not only infertile but utterly uninhabitable.[25]

Faced with catastrophe, Nauru's government made a series of increasingly desperate attempts at self-preservation—alternating between laundering money for the Russian mafia and recognizing breakaway states in return for cash. Eventually, in 2001, Nauru's options ran out and it agreed to become part of Australia's "Pacific Solution." The consequence of this was a stream of aid but its price was hosting a massive migrant-detention facility. By using Nauru as an

22. Che Gossett, "Blackness, Animality, and the Unsovereign," *Verso Books* (blog), September 8, 2015, https://www.versobooks.com/blogs/2228-che-gossett-blackness-animality-and-the-unsovereign.

23. Ben Doherty, "A Short History of Nauru, Australia's Dumping Ground for Refugees," *The Guardian*, August 9, 2016, sec. World news, https://www.theguardian.com/world/2016/aug/10/a-short-history-of-nauru-australias-dumping-ground-for-refugees.

24. Anja Kanngieser, "Climate Change: Nauru's Life on the Frontlines," *The Conversation*, October 21, 2018, http://theconversation.com/climate-change-naurus-life-on-the-frontlines-105219.

25. EJOLT, "Phosphate Mining on Nauru," *Environmental Justice Atlas*, accessed April 25, 2019, https://ejatlas.org/conflict/phosphate-mining-on-nauru.

offshore prison, Australia avoids its responsibilities under the 1951 Convention Relating to the Status of Refugees. Since 2013, Australia has used its offshore prisons on Nauru and Manus Island to prevent anyone who seeks asylum in Australia by boat from landing on its sovereign shores. Nauru, not a signatory to the convention, provides the perfect alibi for the detention, deprivation, abuse, and torture of thousands of racialized migrants.[26]

Nauru is an object lesson between the extractive colonialism and the violence of the border. The island was stripped of its resources because of colonial contempt for the Indigenous islanders. The poverty of those same islanders could then be weaponized to use the island as a site of racialized violence. The population of Nauru, rendered surplus, is forced to deputize the oppression of migrants from other surplus populations. The island sustains itself on the destruction of the lives of others, because all other means of sustenance have been ruined. Nauru's experience encapsulates the breadth and depth of the disaster's destructiveness: an example of what is happening, what has happened, and what is going to happen.

Reading of this breadth and depth, you, like us, might feel the pull of despair. But to despair over Nauru is to return to the same problem we found above—where an understandable despair at the reality of destruction becomes confused with an unacceptable hopelessness at the inevitability of cruelty. In the face of such calamity, do not doubt the capacity for kindness. Even amid the horror of the camps on Nauru, the prisoners gather together and organize for their collective survival. Despite years of police repression, 2016 saw protracted protests by those detained on Nauru and with solidarity from Nauruans.[27]

In these protests we glimpse the beginnings of disaster communism. We are aware that the word "communist" risks conjuring up images of authoritarian statism. Yet there is no name other than "communism," that Out of the Woods knows of, adequate to describe collective world-building beyond the state and capital. For us, communism is precisely the process which simultaneously undoes "business-as-usual" and builds a new world. Communism, in short, is the real movement which abolishes the present state of things. Like ecological crisis, communism cannot be understood as yet-to-come. Communism, too, has existed, still exists, and will continue to exist. This is what provides hope against grief and replaces acceptance with struggle.

---

26. Doherty, "A Short History of Nauru, Australia's Dumping Ground for Refugees."
27. "Protests Escalate on Nauru," *Refugee Action Coalition* (blog), April 6, 2016, http://www.refugeeaction.org.au/?p=4859.

Like disaster, communist struggle is differentiated. Those undertaking it are more often than not already feeling the sharp end of ecological crisis: Indigenous peoples, migrants, racialized people, women, prisoners, "queers," workers, the poor, and the disabled. Isolated, their struggles can appear reactive, as if they provide only temporary local reliefs. Capital is all too eager to offset the costs of its ecological crisis onto those who suffer from it, attempting to turn coterminous struggles into self-ingesting infighting. When viewed together, the acts of these groups appear the prime motor of social change. We know not yet what we might do, and this unknowable togetherness, we call communism.

It is this which makes us hopeful, which wards off that damaging and self-fulfilling despair. Hope is our word for the grave but positive emotion which collectively emerges within the disastrous present, pushes against it, and expands beyond it. With Ernst Bloch, we insist that this hope is not expectation, nor even optimism.[28] Rather, it is always against itself; warding off its tendency to become a fetish, sundered from solidarity and struggle. This is hope against hope.

The importance of being together and becoming together is one we feel strongly about as a collective. Through the simple repetition of talking and writing online, Out of the Woods has become an important part of all our lives, with shared study evolving into real care and solidarity. It has been a wonderful thing to write together: typing over each other in sprawling online documents, not remembering or caring which parts any individual wrote, piecing together our knowledges on things we already knew, teasing out from each other things we didn't know we knew, and collectively addressing those things we did not and do not yet know. As a collection of essays-thus-far written, this book is by no means the culmination of our thought but a series of snapshots of thought-in-gestation. Any kind of conclusive finality is impossible for us. As in the struggles we advocate, this process of becoming together can have no destination at which it settles once and for all. We frequently disagree with each other about what we wrote yesterday, about what we are writing today. This too prevents any sense of finality, as does the fact that, come tomorrow, we want to be writing with each other again. Our thinking together is not complete because it can't be completed, and even if it could, we wouldn't want it to end anyway.

Writing as a collective under a shared name solidifies this becoming together. Yet we recognize that it can also play an obfuscatory role, allowing us to escape accountability for our histories and positions, and eliding our relationships to those

---

28. Ernst Bloch, *The Principle of Hope, Volume 1*, trans. Neville Plaice, Stephen Plaice, and Paul Knight (Cambridge, MA: MIT Press, 1995).

power structures which reproduce the ecological crisis. Out of the Woods started from a call circulated online in English, predominantly shared in a communist milieu concentrated in the UK. The founding members were all loose acquaintances and largely affiliated with UK universities, whether as staff or students. At this point, we were all white and all men. This probably reflected the nature of the call—a (perhaps uninviting) invitation to do unpaid theoretical work, with all the imbrications of privilege that inherently involves. The original composition of the collective was reflected in the readership we appealed to: a certain left-theory audience was implicit in our writing. In the years since, new people have joined Out of the Woods—primarily through Twitter. Others have stepped back. The collective is now spread across the United Kingdom and the United States, and while it is no longer all men and certainly not heterosexual, we are still all white— and several of us settlers in North America. We are undoubtedly beneficiaries of the nexus of raciality, capital, and colonialism. We take responsibility for and fight against being determined by that inheritance. Several of us still work on the margins of universities, as tenuous students, temporary lecturers, and administrators. It is important to keep in mind these situated perspectives as you read this book—not to invalidate our thought but to better specify it. What has (not) been written undoubtedly reflects those who have (not) written it. With this book complete, Out of the Woods will transform yet again, with the intention of further multiplying our positions against homogeneity. We invite you to contact us, and to think, write, and struggle with us.

When we formed Out of the Woods, we wanted to intervene against the consistent inadequacy of many existing narratives around the ecological crisis. We profoundly disagreed with mainstream environmentalism's call for a unified humanity that might stand against a yet-to-come cataclysmic event. Simultaneously, we were appalled by the ways this homogenous conception of humanity coexists with moralizing critiques blaming cataclysm on an excess of humanity. Through such a process, cataclysm becomes too easily pinned on "dirty" developing countries, with their "rapidly" reproducing populations, and their "floods" of migrants.

Oppositions between the polluted and the pure, the populating and the controlled, and the migrating and the placebound all ultimately depend on another organizational divide: between white "civilization" and racialized "disorder." The concept of "nature" serves as an avatar for white anxiety, making manifest fears around the loss of purity and control. Such fears can supposedly be overcome only through the imperial orchestration of intergovernmental organizations—or, a descent into war. The ecological crisis, while supposedly undifferentiated in its effects on humanity, is overdetermined in its causes.

As antiauthoritarian communists and anarchists, we oppose these articulations of environmentalism. However, we also find similar reasons to oppose many of the conventional leftist responses to the ecological crisis. Green anarchism, for example, has too often been focused on defending localized purity of autonomous zones, which in our view constitutes a dangerously introverted response to a globally differentiated disaster. In its romantic attempts to find a pure nature to defend or return to, green anarchism has orchestrated another set of violences.[29] At times, an essentialist conception of nature has been expanded into gender, with some (most infamously Deep Green Resistance) articulating transphobic views. Other anarchists—especially in North America—have used primitivism as an excuse to imitate and appropriate Indigeneity. Shockingly, criticisms of such projects have been labelled as sectarian slurs and intra-anarchist scuffles, rather than substantive disagreements. Unlike the well-circulated 2009 anonymous anarchist text *Desert* or the reactionary Dark Mountain Project, we are uninterested in a project of "nature-loving anarchism," nor do we countenance such works' trendy poetic nihilism disguised as sober realism. "Nature" today emphasizes only a separate and fallen world. It is no real shock that nostalgia for a now-spoilt nature is a frequent theme in reactionary thought, the "romantic anticapitalism" that Iyko Day properly names.[30] Such a racial project, she argues, is premised on both the appropriation of Indigenous lands and practices by settlers *and* the exclusion of those deemed corrupted by capital—Asians, Jews, and migrants more generally. Against romantic anticapitalism, then, we draw inspiration from Black, Indigenous, and anti-border anarchists and communists, for whom romantic anticapitalism can only reek of white supremacy.[31]

This book reflects our desire to write something useful, to create something that makes it easier to understand the total interdependence of the extractions, exploitations, enslavements, and extirpations that colonial capital has brought upon this world. The brutal techniques of these myriad forms of ruination are too often reproduced in green aesthetics, politics, and practices. The logics of reactionary ecology, border imperialism, and racialized state violence are perpetuated and

29. Josie Michelle, "Against the New Vitalism," *New Socialist* (blog), March 10, 2019, https://newsocialist.org.uk/against-the-new-vitalism/.

30. Iyko Day, *Alien Capital: Asian Racialization and the Logic of Settler Colonial Capitalism* (Durham: Duke University Press, 2016).

31. See William C. Anderson and Zoé Samudzi, *As Black as Resistance: Finding the Conditions for Liberation* (Chico, CA: AK Press, 2018); Harsha Walia, *Undoing Border Imperialism* (Oakland: AK Press, 2013).

proliferated in environmentalism. Such a situation demands a different and differentiated response. We see such a response in inchoate tendencies all around us. This book is about the countless ways people survive amidst and against the ruins. We believe these shared strategies to survive ecological crises make a collective thriving within and beyond ruination possible. Thus, in the service of something that truly changes everything—which is to say *planetary revolutions*—we offer new concepts to hold together, and hold close, amidst the continuation of the crisis. Against gleeful doomsaying, romantic anticapitalism, and hopeful technofixes:

> We *hope-against-hope*
> for a careful, yet fierce, *queer cyborg ecology*
> built through a *bricolage* of tools, techniques, and knowledges already
>   around us
> to move *within, against, and beyond* the ecological crisis
> for *survival pending revolution*
> to make, altogether, *disaster communism.*

GUIDE TO THE BOOK

Hope Against Hope: Writings on Ecological Crisis is organized into four sections: **BORDERS, NATURES, FUTURES,** and **STRATEGIES.** Each section is preceded by an introduction that contextualizes the essays in our thought and in the world. We highly recommend reading these introductory pieces, all written in 2019, before the essays contained within. We have updated these essays for internal consistency across the book, albeit admittedly some essays are very much products of the moment in which they emerged. We have, through further citations and introductory remarks, revised the shortcomings of our earlier thoughts. We hope the contradictions that remain can be fruitful for readers. In particular, we continue to work through the contradiction between an understanding of ecological crisis as something that reproduces and is reproduced by other social crises and a more expanded notion of ecological crisis as something that *incorporates* these crises (though does not eradicate their specificities). Although we have discussed this tension amongst ourselves, no clear consensus has yet to emerge (and this is as true for some of us as individuals as much as for the collective as a whole).

We begin with **BORDERS,** the struggles against nationalism, enclosure, and immobility as imperial projects of nation-states. We suggest that these features of our world play a key role in reproducing ecological crisis; and that the latter must not be understood as separate from those punitive regimes seeking to manage human mobility. The interview and four essays in this section forcefully argue for a politics beyond and against the border imperialism of nation-states.

The essays in **NATURES** unpack how understandings of what and where "nature" is affect: the ability of capital to appropriate and extract value. Nature is never as self-evident as it is made to seem; in fact, there is nothing less "natural" than nature. We demonstrate this claim through an evaluation of the role the defense of nature has played for reactionary and fascist individuals and political movements. We further show how the understanding that nature is always *produced* can liberate us to collectively construct better worlds.

Perhaps nothing is made more visible by climate disaster than the manner in which the future is very much at stake in the choices we make today. **FUTURES** contains three essays concerning what this means for climate and left politics. In mainstream environmentalism the future is too often a safe, knowable realm for heterosexual white children to inherit. By thinking through a range of struggles, we reposition the future as an unknowable site of possibility for a variety of subjects in constant formation.

There are many proposals for what is to be done in the face of disaster. **STRATEGIES** outlines our analysis of contemporary leftist approaches to the ecological crisis, especially those which under the name of climate justice somehow seek to obviate or even exclude struggles against capital, coloniality, and racism. But we refuse to remain only negative. In our most recent and comprehensive essay, we outline the ways disaster communism might be understood to be already emergent amidst such environmental crisis.

Much more could be said about these essays and their possible strengths and weaknesses. We remain self-critical, but this book presents an opportunity to expand the scope of critique. As disaster engulfs spaces and times around us, let us put to you a simple question: what will it take to get out of the woods, together?

I

BORDERS

INTRODUCTION

DISASTER MIGRATION

As the ecological crisis accelerates and its effects are exacerbated, people are driven to leave their current places of residence in search of somewhere else. While images of the inhabitants of small-island states forced to abandon coral atolls dominate the popular imagination of climate-induced migration, the reality is more complex. Ecological crisis is not limited to climate change, and environmental factors are often insufficient explanations for migration. In many cases it is only when higher temperatures, rising waters or increased pollution are compounded by issues such as hunger, poverty, a poor quality built environment or warfare that migration becomes a necessity. These issues are, of course, all inter-related with our changing environment, but also have long histories through colonial capitalism's disposessions and enclosures. It is this imbrication of environmental and (ongoing) factors that we refer to as 'ecological crisis.' Most people who move do so within their own countries. In the cold language of international law they are "Internally Displaced Persons." This partly explains why, historically, the issue of ecological migrations has not been of particular concern to the imperial heartlands of North America and Europe.

In recent years, however, the states of the Global North have come to realize the temporary character of this reality. As ecologies destabilize and conditions worsen, many of the places currently serving as refuges will become uninhabitable.

Traveling to the higher-latitude zones, and the richer states that currently patrol and police these spaces, will likely become more essential. Living in these places will not insulate people from disaster, but it will make many of them less vulnerable to disastrous events. This is not least because wealthy nation-states remain better equipped—at least financially—to mitigate such events. However, from the point of view of these states, the prospect of millions of new migrants is already *itself* a disastrous event which must be mitigated. And they are already preparing.

So far, the international response to migration has consisted primarily of a rush to make predictions and distinctions, to quantify the number of people who will move, and to qualify the reasons for their movement. These responses emanate from a desire to measure and manage the growing crisis foretold by these quantifications. The most popular of these predictions has been that of Norman Myers, whose claim that there will be 200 million "environmentally displaced" people by 2050 has been widely repeated. The sociologist Stephen Castles has cast doubt on the accuracy of this prediction, suggesting that Myers' "objective in putting forward these dramatic projections was to really scare public opinion and politicians into taking action on climate change."[1] Such action is not hypothetical; militaries and border patrols are already engaging in preparatory activities and field games in preparation for mass migration.

For Castles, national security is "a very laudable motive," but we are significantly less enthused. After all, it is not ecological crisis per se that necessitates action but the specter of mass migration which, in Castles' words, is deployed to "scare public opinion and politicians." In this connection, Myers rewrites the nature of the threat. Ecological change does not pose a threat to people directly, but produces people who pose a threat (to other people). In the narrative of Myers' prediction, the problem becomes the migrants.

In a paper entitled "Environmental Refugees: An Emergent Security Issue," Myers writes:

> The 1995 estimate of 25 million environmental refugees was cautious and conservative. . . . To repeat a pivotal point: environmental refugees have still to be officially recognized as a problem at all. At the same time, there are limits to host countries' capacity, let alone willingness, to take in outsiders. Immigrant aliens present abundant scope for popular resentment, however unjust this reaction. In the wake of perceived threats to social cohesion and national identity, refugees can become an excuse for outbreaks of ethnic tension and civil disorder, even political upheaval.[2]

1. Quoted in Hannah Barnes, "How Many Climate Migrants Will There Be?," *BBC News*, September 2, 2013, https://www.bbc.com/news/magazine-23899195.
2. Norman Myers, "Environmental Refugees: An Emergent Security Issue" (Report for the 13th Organization

Unsurprisingly, states around the world are far more sympathetic to this formulation of the threat than one that would locate the problem in capitalism. While the last fifty years have demonstrated the inability of states to reduce emissions or adapt to climate change, this history has proved testament to states' increasing interest in, and capacity to, control migration.[3] In other words, if the threat of climate change is posited as mass migration, then the state has already found its solution—the border. The question of who and what will be allowed across the border when and where becomes simply a matter of managerial distinctions and administration.

Myers' numbers never function innocently as a mere prediction of displacement; rather, they necessarily function as a provocation for the prevention of movement. In our view, current attempts within the European Union and the United Nations to forge a definition of what constitutes a "climate refugee" should, by the same token, be seen as a border operation and not an ethical enterprise. The figure of the "real," "deserving" climate refugee will inevitably be deployed against the "undeserving," "ordinary," and "risky" migrant.[4] Liberal *New York Times* columnist Thomas Friedman offers a particularly clear example of decisively drawing such a line. He suggests he has sympathy for "people truly fleeing tyranny" and escaping "climate change, overpopulation and governance stresses fracturing [their] countries." However, "economic migrants gaming the process" must be distinguished, filtered out, and repatriated.[5] Leaving aside, for a moment, the depravity of such an argument, it must be stated that it is more or less functionally impossible to separate one's experience of climate change and global capitalism more generally. The process of quantifying and delineating those who might move contributes precisely to the practice of qualifying *those who can move.* Measurement and definition inspires management.

In other words, the statistics are not just generally shocking, they are engineered to create a very particular form of shock: one that runs along the lines of planetary class and race and culminates in the desire to defend the border. And, unfortunately, the "shock value" of Myers' prediction remains hard for many environmentalists to resist, even when one has demystified the claims as we have just sought to do. In fact, we should add ourselves to the list, as we originally made use of these numbers in the essays that follow! Twice, over the years, we reproduced

for Security and Co-operation in Europe [OSCE]), May 25, 2005. Available at https://www.osce.org/eea/14851.

3. Reece Jones, *Violent Borders: Refugees and the Right to Move* (London and New York: Verso Books, 2016).

4. Giovanni Bettini, "Climate Barbarians at the Gate? A Critique of Apocalyptic Narratives on 'Climate Refugees,'" *Geoforum* 45 (2013): 63–72.

5. Thomas L. Friedman, "Trump Is Wasting Our Immigration Crisis," *The New York Times*, April 25, 2019, sec. Opinion, https://www.nytimes.com/2019/04/23/opinion/trump-immigration-border-wall.html.

Myers' prediction, which is to say, we attempted to turn its shock value to our own ends. Even though those ends are generosity and the destruction of borders, the mobilization of outsized statistical fears was a failure on our part. We no longer feel the provocative power of the number "200 million by 2050" can be legitimized by those who share our politics. As such, Myers' transfixing numbers were a means that contravened and undermined the ends we sought, for, as a statistical incitement to border violence, it can never be replicated in defense of migrants.

We have chosen to leave the numbers in our interview below so as to be accountable for our mistakes. Equally, such an action demonstrates the dangerous allure these numbers hold. They are a reminder of the need for relentless critique, not only of the work produced by others, but of that which we write ourselves. It is clear we must resist both the nativist-racist fear embodied in these predictions and the cognitive border operation inherent in the distinction between "migrant" and "climate refugee."

The essays in this section reflect our collective sense that the differentiated catastrophe of climate change is nowhere more in evidence than in the border practices of states. Climate change makes it increasingly impossible to live in places largely occupied by the racialized, the colonized, and the impoverished. The border seeks to retain or return or to break migrants down enough that they are willing to perform grueling labor at lower rates of pay. Committing to the ongoing struggles against the operation of the border is therefore essential to any practice against climate disaster.

These struggles demonstrate that the border is not confined to the site of the frontier, but rather is a structural part of the nation-state.[6] During the blockades of Immigration and Customs Enforcement (ICE) facilities across the United States in 2018—attempts by activists to disrupt detention and deportation efforts—activists immediately encountered the all-pervasive nature of the border. In Philadelphia, the summer-long Occupy ICE encampment mutated rapidly into a multi-issue movement characterized as "Black-led autonomous revolutionary organizing of the unhoused."[7] This mobilization was capacious, featuring actions in solidarity with Puerto Ricans, people with addictions, nonimmigrant prisoners, and victims of police violence.[8] In the UK, recent organizing has confronted the border in schools, as part of a successful mobilization against the gathering of pupil nationality data.[9]

6. Walia, *Undoing Border Imperialism*.

7. Anonymous Contributor, "This Movement Is Not Ours, It's Everybody's," *It's Going Down* (blog), July 25, 2018, https://itsgoingdown.org/this-movement-is-not-ours-its-everybodys/.

8. Anonymous Contributor, "Occupation, Revolt, Power: The 1st Month of #OccupyICEPHL," *It's Going Down* (blog), August 14, 2018, https://itsgoingdown.org/occupation-revolt-power-the-1st-month-of-occupyicephl/.

9. Against Borders for Children, "We won! DfE are ending the nationality school census!," *Against Borders for Children* (blog), April 10, 2018, schoolsabc.net/2018/04/we-won/.

While this particular state initiative was defeated, the fact that similar practices persist in healthcare, higher education, and housing demonstrates the unconfined reality of border operations.[10]

State and capitalist actors frequently dither about or outright deny the climate crisis. Too often, however, this serves as a useful distraction from the fact that they are, all the while, *actually preparing* for the imminent reality of mass displacement. For instance, states are investing massive amounts of money in technologies that exacerbate existing geospatial inequalities and keep these increasingly unequal populations separate. Ecological dystopia for the many, in other words, could still be utopia for the few. The trend towards global movement north will likely intensify efforts to cordon off these relatively privileged zones: the astonishingly self-described "military-environmental-industrial complex" is already plotting new forms of violence to defend European, North American, and Australian borders and to expand profits.[11]

It is important, however, not to mistake the increasing omnipresence of the border for omnipotence. The reason that the state must constantly attempt to maintain control over borders—and their futile attempt to categorize and separate people, often through violence—is because they are brittle. We must hold onto victories so as to remember the border is not all-powerful and can be abolished. In these essays we return to an example from Glasgow in the nineties, where a buddy scheme partnering recent migrants with locals built bonds of solidarity and kinship.[12] The scheme was such a success that when the state attempted to detain some of the migrants in dawn raids, they found themselves confronting a working-class community united in defense of their friends. Dawn raids ceased.

The bunker-network of the planetary ruling class is by no means a *fait accompli*. Proliferating borders can be—and are every day being—opposed. Communal efforts to combat such violence, such as those in Glasgow, will form some of the most important struggles against ecological disaster. The essays in this section, more than anything, seek to explore border struggles *as* ecological struggles.

---

10. See Docs Not Cops, "#NHS70—No Borders in Healthcare," *Docs Not Cops* (blog), July 5, 2018, http://www.docsnotcops.co.uk/nhs70-no-borders-in-healthcare/; Erica Consterdine, "UK to Remain a Hostile Environment for Immigration under Nebulous New Post-Brexit Policy," *The Conversation*, December 20, 2018, http://theconversation.com/uk-to-remain-a-hostile-environment-for-immigration-under-nebulous-new-post-brexit-policy-109095.

11. Todd Miller, *Storming the Wall: Climate Change, Migration, and Homeland Security* (San Francisco: City Lights Books, 2017), 47.

12. Maryline Baumard, "Give me your tired, your poor . . . the Europeans embracing migrants," *The Guardian*, August 3, 2015, https://www.theguardian.com/world/2015/aug/03/europeans-who-welcome-migrants.

This section begins with an interview with two of our members by *BASE Magazine*. Contextualizing border politics in terms of care, the conversation ranges widely over questions of natures, futures, and strategies. It also serves as an introduction to our conceptualization of borders as a means of differentiating the impacts of climate change and to our thought and politics more generally. Alex tweaks a formulation from the geographer and prison abolitionist Ruth Wilson Gilmore which might serve as the foundation for our politics: climate change is the group-differentiated destruction of the means of our survival.[13]

Alex also draws on the work of the Cameroonian philosopher Achille Mbembe, who has written evocatively about the use of state violence to produce "death-worlds" of populations exposed to violence. This is a theme taken up in the second piece, "Refuges and Death-Worlds," in which we seek to find spaces of survival which can serve as antonyms and antidotes to the ongoing production of disaster. A politics of refuge is, for us, a planetary politics which insists that everyone has a right to a habitable place. Again, to quote Gilmore, "freedom is a place."[14] So, if the habitable zones of the Earth retreat pole-ward without regard for the sovereignty of nation-states, then we must look beyond nation-states and the white supremacist "lifeboat ethics" we identify in Section II of this book.

In "Infrastructure Against Borders," we argue for building webs of mutual aid across borders and migrant/citizen divides. Creating such infrastructures of solidarity over the coming decades is vital for undermining the currently prevailing "build the wall" mentality which has turned the Texas desert and the Mediterranean Sea into mass graves, and the island of Nauru into a detention camp.[15] These are material infrastructures as well as infrastructures of feeling— the "consciousness-foundation, sturdy but not static, that viscerally underlies our capacity to . . . select and reselect liberatory lineages" of ancestors and their capacious and expansive struggles.[16] It is difficult to give a full account of the radical potential of such infrastructural practices. China Medel describes how humanitarian aid in the desert by the direct-action group No More Deaths undergirds an "abolitionist care":

13. Adaptation of Gilmore's definition of racism in *Golden Gulag*, 246.

14. Ruth Wilson Gilmore, "Abolition Geography and the Problem of Innocence," in *Futures of Black Radicalism*, eds. Gaye Theresa Johnson and Alex Lubin (London and New York: Verso Books, 2017), 227.

15. Democracy Now! Staff, "Mass Graves of Immigrants Found in Texas, But State Says No Laws Were Broken," *Democracy Now!*, July 16, 2015, http://www.democracynow.org/2015/7/16/mass_graves_of_immigrants_found_in; Saeed Kamali Dehghan, "Migrant Sea Route to Italy is World's Most Lethal," *The Guardian*, September 10, 2017, sec. World news, https://www.theguardian.com/world/2017/sep/11/migrant-death-toll-rises-after-clampdown-on-east-european-borders.

16. Gilmore, "Abolition Geography and the Problem of Innocence," 237.

In our practices of care, No More Deaths actively works against the neoliberal process of strategic abandonment, in which governing bodies carefully eschew responsibility for a minoritized social group deemed valueless by a logic of racialized criminalization. Sequestered in the Sonoran Desert, the camp wakes up each day committed to practices of taking care, not only of migrants in distress, but also of one another. In the practice of care, desert aid workers prefiguratively build a world in which hierarchies of human value are abolished, where migration is an expression of life making, and where food, shelter, medical, and emotional care are available to all, regardless of notions of deservedness. This care work becomes an abolitionist gesture of direct action that builds alternative forms of recognition and inclusion against the logic of criminalization and the production of valueless life functioning to "protect" the United States.[17]

Such actions actively work against the weaponization of the desert accomplished by the US' "prevention through deterrence" policy, which in many ways, might be seen as analogous to the UK's "hostile environment" policy. Broadening the understanding of environment here, our essay "A Hostile Environment" seeks to demonstrate how contemporary British and American border imperialisms are tied to the maintenance of white supremacy. Having established our analysis of xenophobia in the North as a racist reproductive politics steeped in old fears of sexual defilement and miscegenation, we advance the usefulness of a concept of "critical dystopia" to think though the bleakness of this political moment without foreclosing the prefigurative, even utopian, struggles described by Medel above. This essay has been updated since its original publication. The addition of recent examples more thoroughly flesh out the logics we describe, and have shown their extension in even the eighteen months since the original essay was written.

In order to adequately address climate change, a project that develops a politics beyond and against the border imperialism of nation-states is required. While there are robust border abolitionist movements scattered around the world, much like our examples of disaster communism, these spaces are not yet robust and interconnected. Undoubtedly, this is in part due to the power of military and police forces, which we should not underestimate. Yet there is also a disappointing lack of attention to, and investment in, such struggles from the Green Left. For our part, we insist on the necessity of a "no borders" politics to the ecological crisis. Like communism itself, this is both a movement that abolishes the present and a description of a world beyond that present. It is a necessary condition—perhaps *the* most necessary condition—for a livable world of ecological flourishing.

17. China Medel, "Abolitionist Care in the Militarized Borderlands," *South Atlantic Quarterly* 116, no. 4 (October 1, 2017): 847.

ON CLIMATE/BORDERS/ SURVIVAL/CARE/STRUGGLE

TWO MEMBERS OF OUT OF THE WOODS IN CONVERSATION WITH *BASE MAGAZINE*

First published June 2017

What follows is an edited transcript of a conversation that took place via Skype. We have smoothed over some of the infelicities that result from spontaneous speech to make it easier to read while preserving some of the clunkiness of phrasing to convey the texture of spoken dialogue. Footnotes attend to some of the mistakes made in our arguments here.

It feels particularly fitting that this interview should be the opening piece in this book. Among the digressions and mistakes, this conversation is also full of new ideas, some of which have subsequently become fundamental to our thinking. This is the first time we talk about the false image of a spectacular, singular apocalypse; the first time we define (after Gilmore) ecological crisis as the group-differentiated destruction of the means of survival; the first time we use the term "catastrophic present." Indeed, a strong theme in this interview

is differentiation: that the disasters people experience, the struggles they organize and the futures they struggle for are all differentiated by race, class, gender, and sexuality. Black feminist critiques of universalism and humanism exert a strong influence on the entirety of this interview, but particularly on the sections pertaining to migration and borders. It is fitting that this book begins with a generative conversation, because that is where all our work begins. Conversations amongst ourselves, amidst the work of others, is the ulimate origin of all our thought. That now, on reflection, we can see mistakes and mishaps here, only serves to demonstrate the nature of collective study; that we know differently now to how we knew then. In particular, our ongoing theorizing of 'ecological crisis' as something connecting environmental concerns with the violences of borders, prisons and racialization, grows out of some of what we discuss here.

BASE Magazine: In much of your writing, you talk about the relationship between mass migration and climate change. How can climate change be more consciously linked to existing opposition to borders and everyday struggle against the border regime?

Out of the Woods, A: One place to start would be the estimate of 200 million climate migrants by 2050, which Norman Myers put out over a decade ago. This is seen by many as a conservative projection, yet even so, it would mean that by 2050 one in every forty-five people in the world would have been displaced by climate change.[1] A report for the International Organization for Migration notes that, "on current trends, the capacity of large parts of the world to provide food, water and shelter for human populations will be compromised by climate change."[2] The framing of this "capacity" as a series of absolute, "natural" limits is of course problematic. "Carrying capacity" is a product of racial heteropatriarchal capital as it works through nature and of nature as it works through racial heteropatriarchal capital. However, climate change *will* certainly erode people's capacity to reproduce themselves and in a manner that forces their movement. The majority of climate migrants will be racialized people, and it seems highly unlikely that those states least affected by climate change and/or most able to adapt to it (the white powers of Europe and America), will approach climate migrants any differently to those racialized people already being murdered by their borders or imprisoned in their camps. Climate change is another reason people have to move, but it is not a reason for states to treat moving, racialized people any differently.

Out of the Woods, D: When Black Lives Matter UK shut down London City Airport they were very clear in stating that climate crisis is racist. It disproportionately affects people of color, both because they can't cross borders with the ease that white people do—for a whole host of reasons—and they're more likely to live in areas that are worst affected by climate change. Connecting up struggles that might be seen as "single issue" in this sense is really important because, in a sense, they *are* single issue: climate change and racism reproduce each other.

BASE Magazine: Since it features heavily already, and will likely appear again, could you speak a little more to the nature of the border—its composition and politics?

---

1. See our introduction to Section I, "Introduction: Disaster Migration," for a critique of our use of Myers' numbers in this interview.

2. Oli Brown, "Migration and Climate Change," *International Organization for Migration Research Series* (Geneva: International Organization of Migration, 2008), 17. Available at https://www.iom.cz/files/Migration_and_Climate_Change_-_IOM_Migration_Research_Series_No_31.pdf.

D: The violence of the border isn't just *at* "the border"—schools become borders, hospitals become borders. I broke my knee recently, and I—a white person who speaks English as their first language—was very well looked after at the hospital [in Nottingham, England]. However, a South Asian woman who came in a few minutes after me didn't fare so well. Her English wasn't great, she wasn't able to think clearly because of the pain she was in, and staff were insisting she gave an address—and she didn't understand what they were saying. Although she did eventually receive care, we know that the NHS will withhold treatment: this is a form of border violence.[3] So, struggles that might seem quite distant from ecological issues—hospital workers resisting the imperative to behave in this sort of way, for example—are really important for a transformative ecological politics.

A: I think when it comes to climate change what we're seeing is the way the border can be used to trap someone within an increasingly catastrophic present. Achille Mbembe has written extensively about *necropolitics*, of holding people within a situation where their life is defined by their proximity to death.[4] The border keeps people in places where they cannot find food or [are] at the mercy of floods. This is coercive, conscious violence orchestrated by states that will persist, both in countries outside Europe and within it. I think we must also emphasize that there's a globalized institution of antiblackness, and the forms of violence which reproduce it are very much in common. The necropolitical obviously operates against Black people in the United States and the UK, as well as in Libya and the Mediterranean. In terms of the way climate change and natural disasters might interact with this existing necropolitics, it is perhaps important to think of police operations in New Orleans in the wake of Hurricane Katrina. On Danziger Bridge, seven police officers opened fire on a group of Black people attempting to flee the flooded city, killing two of them and seriously injuring four more. That event—Black people being murdered by the state—encapsulates the necropolitical violence of attempting to hold people, and particularly Black people, in a place where life is untenable, and then extinguishing that life as soon as anyone tries to move out of that place. That's the murderous double bind of anti-Black violence in the policing of crisis.

---

3. The National Health Service is the UK's public health provider. Famously a "universal" health system feted for providing care on the basis of need rather than ability to pay, the UK government has in recent years developed a range of policies restricting access as part of a "hostile environment" directed at undocumented migrants—including a requirement that people pay upfront for secondary care if they cannot prove their "eligibility" for NHS treatment based on migration status. The group Docs not Cops are challenging this: see page 68 below.

4. Achille Mbembe, "Necropolitics," trans. Libby Meintjes, *Public Culture* 15, no. 1 (2003): 11–40.

D: I also think it's really important that we challenge environmentalism's history and ongoing complicity with racism (and outright white supremacy)—[in which it argues] for closed borders, population control, and sterilization, for example. We've recently had prominent members of the Green Party of England and Wales arguing for reductions in migration in the name of the environment and a "sustainable economy."[5] There was a Paul Kingsnorth essay in *The Guardian* a couple of months ago that's abhorrent; it repeats so many of these tropes.[6]

BASE Magazine: Most of us know very little about climate science, and whilst a great many people work very hard to translate an overwhelming amount of data and fieldwork into accessible writing, the point where trends and patterns meet the daily effects of climate change can feel elusive. Is there more that could be done to orientate the energies of existing struggles and how far into the future should we be looking? To what extent, to take just a single example, should a housing movement engaged in a project to defend access to housing across London take into consideration that it could soon find itself underwater?

D: We often understand climate change as leading to a spectacular future event and this is often understood visually: imaginaries of ruined, flooded, and depopulated cities are really common. But I think this is flawed: it suggests climate change is heading towards a singular "event" that is going to happen rather than something that is already happening, often in less visually perceptible forms. It becomes harder to grow certain crops, for example, and food becomes more expensive. That drives both migration and conflict. Climate change has undoubtedly played a role in the Syrian Civil War.[7]

So, it's wrong on an empirical level to figure climate change as this thing that will happen in the future, but I think it's also unhelpful politically, because that kind of future threat I don't think works as a sufficiently motivating force to affect things in the present. I think, like you say, it can be disempowering. That parsing of climate change as a spectacular future event affects how we behave politically as well, leading to a kind of fatalism whereby people just accept these things. I actually think they empower a certain white, male, heterosexual subject too: they can

5. From memory, I (D) had an essay by Rupert Read in mind here. The truth is even more unnerving, because Read had made such arguments as an "expert" to a UK Parliamentary Select Committee. See publications. parliament.uk/pa/cm201012/cmselect/cmenergy/writev/consumpt/consumption.pdf.

6. See "Lies of the Land" in Section II of this volume for our response to this essay.

7. This is not entirely clear to us today, as argued by Jan Selby et al., "Climate Change and the Syrian Civil War Revisited," *Political Geography* 60 (2017): 232–44. The authors dispute existing evidence of both drought-induced migration and migration-induced conflict.

project themselves into that catastrophe thinking they can start anew—the sort of "cozy catastrophism" that the novelist John Wyndham was (perhaps a little unfairly) accused of. You know—"Oh well, all the poor people have died, but we can have a jolly nice time with our new community on the Isle of Wight."

Public mistrust of experts is also a huge problem because the people we usually hear talking about climate change in the media fall into this category. I think a lot of that hostility is entirely understandable, but rather than get rid of "expertise" in favor of a broad cynical fatalism, we need to think how we can expand the category of expertise and popularize it. We need to amplify the voices of those who live and struggle where climate change meets everyday life: migrants who've moved because they can't afford to buy food; people who've worked the land and seen how changes in climate affect crop growth. They, too, are experts.

BASE Magazine: If these kind of analyses of disasters rooted in a distant future can instead give rise to a paralysis and fatalism, whereby with a long enough timescale, all activities become regarded as irrelevant and inconsequential, how then can these feelings be combatted or even harnessed?

D: It's not necessarily the timescale that's the problem here, or that talking about the future is inherently wrong, but the function of thinking about the future. There is a difference between prediction and extrapolation. Beyond identifying broad trends that are highly likely and factoring them into our thinking as appropriate, I think prediction is really damaging: firstly because we know not to trust it, and secondly because it doesn't leave room for agency. We all know that past futurologies, optimistic and pessimistic, religious, apocalyptic have all been terrible. They're lots of fun, with the capacity to fascinate—we've all enjoyed images from the sixties of the year 2000 full of flying cars—and people have long clung to predictions about the imminent collapse of the global population. But they're just wrong and I think damaging to any attempt to challenge climate change.

I think extrapolation is different; it's the mode of a lot of science fiction. Here, I'm reminded of the claim made by adrienne maree brown and Walidah Imarisha in their introduction to *Octavia's Brood*—a collection of short stories from people of color involved in social justice movements in North America/Turtle Island— that all organizing is science fiction.[8] Perhaps we could think about dystopian fiction here. It's had quite a bit of press recently, but the way much of this is framed

---

8. Walidah Imarisha and adrienne maree brown, eds., *Octavia's Brood: Science Fiction Stories from Social Justice Movements* (Oakland: AK Press, 2015), 3.

is unhelpful, I think. Dystopian fiction is positioned as something that can help us "understand" the present in a narrowly empirical way (which denies agency), and the novels celebrated—*Nineteen Eighty-Four, Brave New World, Fahrenheit 451, The Handmaid's Tale*—are limited even in that sense because they disavow the role that race, in particular, plays in structuring our present. And, in the first three of these, the "victim" is understood to be the abstract individual rather than collective subjects co-constituted by race, class, gender, sexuality, and (dis)ability.[9]

So, instead, I think we need to engage with dystopian fiction that extrapolates from the white supremacist, able-bodied, colonial, heteropatriarchy that structures our world—here I'm thinking of writers like Octavia Butler, Stephen Graham Jones, and Marge Piercy. This isn't just a descriptive process—extrapolation doesn't simply describe our world or even where it's going, but at its best gives us the opportunity to intervene in that through collective struggle. It tells readers that acting in the present *can* make a difference to the future. The science fiction scholar Tom Moylan talks about what he calls "critical dystopias," and I think they're particularly useful here, because they present collective organization and struggle *within* the dystopian society being depicted as well. Even if things continue to get worse, this won't be the end; there is always room for collective struggle.[10]

Having said that, I am a little skeptical about the power of literature, partly because we don't generally read it together anymore—unless you're part of a reading group or reading at a university, you probably read fiction as an isolated individual. I think the popularity of *Octavia's Brood* is interesting: it's got a large social media following, has been used by reading groups, and seems to have opened up a space for collective discussion about the future and how acting now can alter it.[11] It doesn't necessarily have to be literary fiction that plays this role: "design fiction" is a potentially powerful tool too, for example.

9. On the race-erasive, white-feminist universalism of the latter text, particularly in its iteration as a Hulu TV show, see Sophie Lewis: "In Gilead, Atwood's fictional setting, human sexuation is neatly dimorphic and cisgendered—but that is apparently not what's meant to be dystopian about it. It's the 'surrogacy'. . . . [As such] *The Handmaid's Tale* neatly reproduces a wishful scenario at least as old as feminism itself. Cisgender womanhood, united without regard to class, race or colonialism, can blame all its woes on evil religious fundamentalists with guns." [Sophie Lewis, *Full Surrogacy Now: Feminism Against Family* (London and New York: Verso Books, 2019), 10.]

10. For Moylan, "critical dystopia" names a historically specific genre of science fiction arising around the birth of neoliberalism. We're prepared to expand the concept to describe our present, however. [Raffaella Baccolini and Tom Moylan, *Scraps of the Untainted Sky: Science Fiction, Utopia, Dystopia* (Boulder, CO: Westview Press). For more on the use of "critical dystopia" as a descriptive term for the present, see David M. Bell, *Rethinking Utopia: Place, Power, Affect* (London: Routledge, 2017), 20–51.

11. The "Prospecting Futures" research conducted by Lisa Garforth, Amy C. Chambers, and Miranda Iossifidis at Newcastle University has been exploring this issue in relation to online science-fiction reading groups (whose texts have included works by Octavia Butler).

BASE Magazine: *In an older issue of* The Occupied Times, *we asked Silvia Federici about surviving apocalypse(s). She told us:*

> *The prospect of annihilation is a relative one. For many communities in the US—Black communities whose children are murdered by the police in the street, Indigenous communities like the Navajo that have to coexist with uranium mining, communities where unemployment is skyrocketing and the list goes on—apocalypse is now. In this context, we struggle for justice by refusing to separate the struggle against the destruction of the environment from the struggle against prisons, war, exploitation. You cannot worry about climate change if your life's in danger every day, as is the case for so many people in this country.*[12]

What do you recognize in these descriptions as possible points of engagement to building our capabilities to survive?

A: I think it's very interesting how Federici responds to that, and I think part it is in the way that you worded the question: "The consequences of climate change are forcing humanity to contemplate its own destruction in ways it hasn't since the proliferation of nuclear weapons at the height of the Cold War." I think what that comes back to is what we were saying earlier about these images of universal catastrophe. Because that question very much sums up the way climate change is depicted in terms of this global, universal threat to the species and a particular framing of the human, but I think it's very important to then pose the questions that Black studies has insisted on: Who is the human? Who gets to be human? Sylvia Wynter's work here is incredibly important.[13]

I think what climate change actually requires us to do is to recognize that it's not one apocalypse. What's more terrifying to think about, but is perhaps more useful, is to realize that catastrophe and normality can coexist quite happily; that it's not about some apocalyptic future but a catastrophic present. This seems especially pertinent in the situation where we had 5,000 migrants drowning in the Mediterranean last year, yet the current discussion is around the fiscal effects of Brexit. There is no squaring of that circle. In reality, Europe is experiencing a form of normality at the moment which is in complete contradiction to these catastrophes. I think what that question requires us to do, and what Federici starts to approach in

12. Silvia Federici, "Preoccupying," *The Occupied Times* (blog), October 25, 2014, https://theoccupiedtimes.org/?p=13482.

13. Sylvia Wynter and Katherine McKittrick, "Unparalleled Catastrophe for Our Species? Or, to Give Humanness a Different Future: Conversations," in Katherine McKittrick, ed., *Sylvia Wynter: On Being Human as Praxis* (Durham: Duke University Press, 2015), 9–89.

her answer, is a differentiated vulnerability and the fact that catastrophe has always existed for some people.

However, I don't think I can agree with her saying you cannot worry about climate change if your life's in danger every day, because I think the people who've been historically struggling against that vulnerability were the first people to experience climate change. The people who've been displaced in Bangladesh, the Navajo Nation, the Standing Rock Sioux (who're fighting the development of the Keystone Pipeline), I think those people have historical experiences, perhaps not always of climate change, but certainly of environmental destruction. When we think about the systematic and organized destruction of the ecosystems of the American Plains and the effect that had on the Indigenous peoples living there, you could say the Standing Rock Sioux have a historical experience of the destruction of the means to survive, not unconsciously as is happening with climate change but very deliberately and consciously.[14] I think what's important to say is that climate change is not unique in its destruction of one's means for survival. To frame it in terms Federici might do herself, it's all about the means by which we reproduce our daily lives. Climate change is the group-differentiated destruction of the means of our survival. Sometimes, for some people, that's going to be catastrophic—meaning the complete obliteration of the means to reproduce yourself—but for others it will be minimal.

That's exactly what we speak to when we talk of these false images of London underwater. One of the things that's so cloying and disgusting about those images is the idea that climate change is a universal problem. What is perhaps more nightmarish about climate change is that it's not; it's a very particularized series of problems that will very differently affect a rich white man who owns a house in Primrose Hill and a Black working-class mother who lives on the floodplains of the Thames.[15] I think there is an important distinction to embrace there. I think that's almost the moment when we must begin to talk of building our capabilities to survive against group-differentiated vulnerabilities. What that forces us to comprehend is the capacity to organize ways to survive.

14. See especially Nick Estes, *Our History Is the Future: Standing Rock Versus the Dakota Access Pipeline, and the Long Tradition of Indigenous Resistance* (London and New York: Verso Books, 2019).

15. This is a bit of a rhetorical simplification—differentiated vulnerability can also mean rich white people choosing to live in more risky places, displacing those who can longer afford to live there. In many places, urban waterfronts are caught between two trends—the increasing desirability of waterfront properties and the exploding costs of living in evermore floodprone areas. Red Hook, a neighborhood built on a peninsula in the floodplains of Brooklyn, has seen accelerating gentrification in the wake of Hurricane Sandy, despite reports that normal high tides will be flooding its streets by 2080. [See Anna-Sofia Berner, "Red Hook: The Hip New York Enclave Caught Between Gentrification and Climate Change," *The Guardian*, September 25, 2018, sec. Environment, https://www.theguardian.com/environment/2018/sep/25/red-hook-climate-change-floodplain-hurricane-sandy-gentrification.]

What I think Federici mentions is the fact that people have always been surviving catastrophes. Here, Out of the Woods would probably talk about disaster communism. Historically, after earthquakes, volcanoes or other moments of instability and damage, people will often exhibit mutual aid, social care, an elaboration of reproductive labor towards liberation. These actions are not contained in, or constrained by, the boundaries of colonial capital or heteropatriarchal individualism. I guess what I'm trying to say is, what Federici gestures towards when she talks of those things like the struggle of Black communities against the police, the struggle of the Navajo against uranium mining, is what Fanon would describe as a program of "total disorder."[16] I guess what we have to think about in terms of resisting climate change, is resistance not just to that but also to the systems of order that differentiate violences.

So, we have to think about organizing against climate change as mediated through a world dominated by colonial, heteropatriarchal capital. The violence is organized and differentiated by these structures and it is in the struggle to destroy those structures that we might also survive. It seems quite evident to me that we can realize a particular imagination that has always been practiced in struggles against catastrophe—struggles founded on care, reproduction, and warmth.[17] Those have always been the things which have made it possible to survive every catastrophe of the past 2,000 years. People will still be fighting those battles even if white environmentalism does nothing about it—that's another thing to insist on. This resistance will happen anyway, no matter what transpires in the corridors of power. It's to what extent we can help each other to go beyond the survival of a few people and emerge from the current series of catastrophes into a world in which we would hope no one experiences them. A world beyond catastrophe is possible.

BASE Magazine: Disaster communism is a concept we've featured in older publications and we've been talking about here again, but it seems that the manner in which it is evoked often relies on the kind of grand "event" which was warned against earlier—for instance, the organizing in the wake of Hurricane Sandy is often brought up as an example of disaster communism in action. The description of care and survival just mentioned now seems to be a far neater deployment of the idea—and that feels a very comfortable fit to the organizing many of us who produce this publication are familiar with (for example, the struggles against the housing crisis and abusive

16. Frantz Fanon, *The Wretched of the Earth*, trans. Richard Philcox (New York: Grove Press, 2004), 2.
17. Automnia, "Ecstasy & Warmth," *The Occupied Times* (blog), August 20, 2015, https://theoccupiedtimes.org/?p=14010.

components of our own social movements). Could we talk more about how if the catastrophe is now, how we may survive it?

A: I've been thinking about disaster communism in terms of what Fred Moten writes about as "fugitive planning": this operation that's always going on beneath the surface of social life because it's the precondition of social life; it's the means of a certain form of collective living.[18] This is familiar for anyone who has had any experience with childcare—there are certain points when someone else looks after your kid whilst you go to the shops or something, and it's a moment which has to happen to make it possible for you to carry out any basic tasks. I guess what's confusing about the way we've been thinking about disaster communism is that there's an uncertainty or vagueness about whether we are calling for something to come into being, or whether we are observing something that's already happening and merely recognizing a certain way of extrapolating it. I think the complexity is that we do kind of use it both ways.

D: There's a distinction between the two modes. There's the "communizing" stuff that's already happening that we can observe, like the kinds of communities that form around disasters, collective relations of care, mutual aid, etc. Then, there's the idea that the term "communism" also names the linking of those struggles on a much larger scale. So communism-as-movement connects these otherwise isolated communizing practices that can actually help reinforce capitalism because capitalism will coopt the common: thanks for self-organizing all this, now we don't have to pay anyone to do it! Also, you've helped increase property values in the area!

A: I guess that's why I was thinking about Moten and planning because, as Moten is saying, against planning there is always policy—the attempt to extract value from planning, to strip-mine the social commons. So all those forms of reproductive labor can easily be exploited by an increasingly desperate state or state-capital formation. This is really notable in frontline care in terms of people being discharged from the NHS early on the expectation that their family will look after them. The policy formation of the state has turned towards care in the NHS being home-based rather than hospital-based, which is in no small part a cloak for the incorporation of planning into policy, and the subsumption of a certain form of social life into the antithesis of that—state and capital. So, I guess this is the ambiguity; what

18. Stefano Harney and Fred Moten, *The Undercommons: Fugitive Planning & Black Study* (Wivenhoe, New York, Port Watson: Minor Compositions, 2013).

already exists wouldn't necessarily destroy the thing that we want to destroy, that's the problem. And this is always the ambiguity of survival as well, you know, survival in a world that depends on your reproduction and your destruction or in holding you in some kind of ground between the two, and that's massively differentiated by race, gender, class, and sexuality. I suppose what we have to do is survive in a way that's antithetical to the survival of the forms of power that oppress us. I guess this is the ambiguity at the heart of disaster communism: how do we survive the disaster whilst also destroying the things that make it a disaster in the first place? How do we become potent whilst rendering the threats to our lives impotent? This kind of constant contradiction or ambiguity is very hard to resolve in theory, but I think can often play itself out in practice.

To bring all of this back to climate change, I think this is what I disagree with fundamentally about Federici saying "you can't worry about climate change if you are already struggling with the everyday," is that it doesn't actually take someone very long to realize that the destruction of their everyday life is based on something bigger than that. People tend to start looking for a pattern, and I think that's the point at which disaster communism has to intervene and say that we can operate on the basis of a destruction of the things that are destroying us.

D: Yes. To say "yes" to what we want—and what is already created in cramped spaces—necessitates saying "no" to the world that dominates save for those cracks or openings. I actually have a slight concern about the phrase "disaster communism" though, which is partly to do with it being such a snappy phrase. I worry that it can travel without the meaning we're trying to outline here, because when you hear "disaster communism" it can bring to mind a communist take on Wyndham's "cozy catastrophism." Like, "hey, if the world ends, we can build a kind of communism."

A: I would agree. I'd probably also go as far as to say that we should try to develop something else because I'm not even sure "disaster" is quite the right kind of word for encapsulating what we are really trying to resist and survive given that it's not one disaster or even a series of disasters, it's a particularly potent mix of catastrophe and normality in which both are murderous. Perhaps the problem of coupling "disaster" and "communism" is that it implies a unified response to a unified crisis, when in fact we have different resistances, necessitated by a group-differentiated schism of normality and catastrophes.

I think the undercurrent to this conversation is the specter of what is now quite openly and explicitly called fascism. We have talked about the potentialities of such fascism in the works of Paul Kingsnorth, and early on in relation to Garrett Hardin's "Lifeboat Ethics,"[19] and how it would be quite easy to imagine a response to climate change in which those at the top of systems of oppressive power, those empowered by capital, the state, gender, class, race, sexuality, basically live out a sort of super-privileged version of what Rebecca Solnit is talking about. The core vision of dystopian films recently has been that either the rich people go and live in the sky or a magic island, etc., but that doesn't seem realistic. Actually, what's more likely to happen is that the city breaks up into increasingly small fragments in which extreme privilege and protected privilege are surrounded by a mass of those who don't have the power to defend themselves, and that plays out around moments of disasters as well. There's several accounts I remember reading after Hurricane Sandy of people watching the streets of New York, just as the hurricane was about to hit, filled with carloads of rich white New Yorkers going to the countryside or to stay in hotels—they were being filmed by Black and Latinx workers who had to stay at work. There's something strong there about the nature of the disaster—some people literally in the absurd, nightmarish situation of not being able to escape the disaster because their boss wouldn't let them.

BASE Magazine: As well as signing a raft of Presidential Memoranda and Executive Orders which reduce the scope of environmental protection oversight for "high-priority" infrastructure and energy projects, the Trump administration also imposed a gag order on offices within the US Department of Agriculture and the Environmental Protection Agency (EPA) to stop them from releasing public-facing documents. In response to the order, and freezes on resources, we've seen dissenting voices from authorities previously in alignment with the state—the National Park Service (@NotAltWorld), the EPA (@UngaggedEPA), and NASA (@RogueNASA) calling for people to #resist. In terms of media dissemination, do these alt formulations present any hint towards a valuable affordance or is the gesture, at best, a populist gimmick? Is there some unrealized application for the alternative channel from the authoritative (peer-reviewed) voice when it comes to information around climate change?

A: I think the reaction to these accounts, and the supposedly "dissenting" elements of the US state they represent, is dangerous to be honest. Celebrating these

---

19. See "The Dangers of Reactionary Ecology" in Section II of this book.

accounts overlooks a lot of the fundamental problematics we have to engage with, and creates a fictional division between some form of rational, proper, scientific state and an irrational, improper, populist perversion of the state.

I think that's dangerous because it occludes a lot of the actual features of the American state which make it so lethal and which are responsible for the current series of differentiated catastrophes which people are experiencing. For example, it's weird to see the National Park Service (NPS) become this embodiment of American liberalism given that the NPS is literally a protracted celebration of a form of wilderness made possible by genocide against Indigenous people. Then again, perhaps it's a good icon for the liberal resistance, because the NPS sets out to preserve a certain kind of pristine purity from the devastation of modernity embodied by urban life (and its associations with blackness). It's actually a colonial myth very similar to American liberalism itself.

I think you can also say related things about the Rogue NASA (*@RogueNASA*) Twitter account. As part of the military-industrial complex, NASA's history and its self-mythologizing as a colonial "explorer" makes it a depressing, if unsurprising, hero for the liberal #resistance. I guess that's what I felt was dangerous about that particular moment in which people started fetishizing a certain form of civil-service resistance. It occludes the nature of the American state and I think we should be careful not to allow populism, or Trump's form of populism, to distract us from the nature of the American state as an organization of forces of heteropatriarchal, settler colonial, capitalist domination—that whole murderous configuration shouldn't be overlooked just because some civil service people don't like Trump.

D: The one thing I would say is that it remains to be seen what kind of forms these movements will take, and certainly in the March for Science there was a lot of very unhelpful exceptionalism —"we are scientists, we produce truth"—which kind of suggests that as scientists they should be protected. In this sense, they failed to join up with already existing struggles and with other movements because they even exceptionalize themselves in relation to other movements. That's worrying, but I'm sure there are elements that do want to connect and do want to join up and are doing so. The NPS, of course, is massively colonial: it has literally forced people off their land and continues to do so. But there may be people who work for the NPS who would like to address this, are aware of this, and would like to remedy it in some way. Just because they are struggling at the moment as rogue employees of the NPS doesn't necessarily mean that they are struggling for a return to what was—you can struggle against your own history as well. Whether that is happening or not, I don't know.

We certainly saw it in the student movement around tuition fees and privatization of higher education in the UK—that wasn't just a struggle for the return of the university as a space where relatively privileged people could have a free education or even be paid to have an education, at its best it was a struggle for a fundamentally different kind of education. So perhaps those struggles will take that kind of direction. I'm sure elements of those struggles will and they are the ones I guess that will potentially have the most interest for radical politics, against and beyond the world as it is.

A: And I guess this is the point where it might be important to talk about a certain form of "treachery" against the manifestation of power that one is willingly and/or unwillingly incorporated into. I have been thinking a lot about treachery in the context of recent discussions around the term "ally." I mean, I think a lot of people have come to realize that the term ally is problematic, but there seems to have been an easy shift towards "accomplice" instead and I don't think that has actually resolved the fundamental problem: both imply that there is some form of easy movement that one can make towards someone in quite a different structural position, which means that you can then unilaterally declare "okay, I am an accomplice now."

I think what this often means, especially for people like myself who are in a particularly privileged position, is that I have to actually think about what it means to be traitorous or treacherous. I think the interesting thing about the figure of the traitor is that you never fully escape the thing that you are betraying. The traitor is always an ambiguous figure who can never be fully trusted because they can always be drawn back into the form they are betraying. So, I guess there is something interesting to think about in terms of these state workers. You know, whenever the police commit another atrocity, they usually pull out some policeman who has a critique of the police, but it never goes into a full betrayal of the police. It's never treacherous, it's always restrained in some way, and I guess it's at that point when you're willing to start comprehending the abolition of yourself, that you might become a useful traitor rather than a very dangerous ally who just seeks to incorporate a more critical edge into the reproduction of violence.

So, there is something about treachery and a willingness to be dangerous to the thing that reproduces you—simplistically speaking, to bite the hand that feeds you. I think if those rogue accounts do become dangerous, it will be if they leak things, if they cause a problem and then are willing to go beyond that. What would be interesting is if those people doing the rogue stuff started quietly talking to and helping Indigenous people reoccupy parts of the National Park Service that have been stolen from them. Maybe that would be a good form of treachery.

BASE Magazine: When it comes to activities to support and build on, people often point to the numerous struggles, many on Indigenous/First Nations land, aimed at preventing the extraction of resources which directly lead to climate change—but much of this seems far beyond the reach of this island. Meanwhile, similar UK-based activity around antifracking seems also to have been rooted in a reactionary nationalism—somewhere between NIMBYism and a defense of the English countryside. How might we better confront and resist the causes and effects of climate change or, if the determining moments are to be far from these shores, how might we better offer solidarity?

A: Once when OOTW were doing a talk, someone from the audience raised this point about Indigenous struggles and was like "we've seen these Indigenous struggles elsewhere and they are really good, important, and fundamental to any kind of environmental practice in the twenty-first century." Which was cool, but then he went "so, what do we need to do in the UK? We need to do something around our local places, our local environments, do we need to become Indigenous?" And that's the moment when you are like "Noooo!" It's ridiculous, but you can see this kind of thing comes up often in the Kingsnorth stuff. It is obviously a real problem and it's interesting because it seems to spread across the political continuum.

Kingsnorth is actually a properly dangerous ideologue who has all of these ideas which have been coalescing around a very fervent nationalism-fascism complex. What's dangerous is that it has been taken up by a lot of people on a liberal Left, who nevertheless seem to find something in it. So, I think part of the problem is that people start making easy equations with the land and start thinking about things in terms of "Nature." What we have always been trying to insist on in OOTW is that there is not some kind of pure nature to go back to, and that any implication of some kind of perfect wilderness is colonial dreaming, and a dreaming which will only vivify an incredibly dangerous form of enclosure of the wild as a means of preserving the world. And, what we've been talking about more in OOTW is the cyborg ecology or the cyborg Earth, in which there is no perfect nature to go back to; and in which we have to face up to the complexities of the interrelation between human and nonhuman life—which ironically enough, is exactly what Kingsnorth says he is trying to resolve! Kingsnorth says he is doing it through the nation, but he can't talk about human and nonhuman life without pitching nonhuman life as some kind of perfect and pure thing. As soon as human life is removed from that, for Kingsnorth, it becomes dirty, polluted, and corrupted because, for Kingsnorth, nature is what rejuvenates the nation.

The thing that we have to resist here is Western colonial romanticism—this absolutely has to be destroyed, and this isn't some kind of abstract literary problem, it totally vivifies a vast proportion of the UK environmental movement at the moment. There is still a popular imaginary of some kind of pure nature which you find as much in RSPB (Royal Society for the Protection of Birds) members as you do in hardcore environmentalist activists, and it really must be refused. At the same time, we need to be certain that we don't become technofuturists who'll happily embrace a technological invasion of everything existing, with no regard for the colonial paradigm and the advent of European technology as both weapon and arbiter of colonial "progress." To a certain extent, we are between a rock and a hard place here—between a romance of wilderness and a romance of technology, and both are worse.

D: I think that binary is really important and you get it from both sides. So, if you try and criticize the fetishization of the slow, the local, the authentic and the romanticization of nature, then you are accused of being in love with the global, the fast or of being a technological fetishist. It's this kind of binary thinking that structures both the accelerationist-oriented, technofuturist Left, and "back to nature" leftism. I think unpicking that binary, in fact rejecting it as a structure, is really important. There is a case, sometimes, for organic food, there is also a case, sometimes, for using drones in farming. And sometimes there is a good case not to grow organic food—we talk about this in our piece on cyborg agroecology.[20] Indigenous ways of organizing life in specific locations across the globe are important here—not so that we can apply them to a wholly different context, but because they often completely undercut those binaries—they are "local," but have dynamic, relational understandings of "local" or "place" that eschew cozy romanticism.

On the appropriation of the term Indigenous outside of the Indigenous context, it's important to be clear that there is no substantive Indigenous population in Britain. I know some Crofters in the Scottish Islands and Highlands argue that crofting is an Indigenous way of life. I don't know enough to comment on that, but generally the way "Indigenous" is used in the political discourse of the UK is to suggest that white British people are the Indigenous population of this island and so have a unique claim to live here. This is sometimes extended "greenwards," so they are held to have a unique ability or right to cultivate its environment, or protect it from "overpopulation."

20. See Section II (NATURES) in this book.

Against that, I would (cautiously) take Indigeneity as a way of naming a particular co-constitution of identity with land and place: a way of life that cannot be separated from the dynamic, relational ecologies in which it developed, and that includes nonhuman life: animals, minerals, and the land itself (and as I understand it, many, though not all, Indigenous people make use of this relational understanding in organizing their struggles). Now if you colonize that land, that way of life is marginalized or made impossible, and that simply does not happen in the UK— left-leaning localists might point to Tesco coming into your high street and closing your local shop. That might be bad, but it's not remotely comparable: your way of life is still fundamentally the same. So the term "Indigenous" just doesn't translate.

I also think there is still a danger of white-settler activists; or white activists in Europe or Britain—and it's a tendency I recognize in myself—fetishizing Indigenous struggles and placing too much hope in them, or just abstracting bits of knowledge without attending to the need for decolonization as a political project. We saw it with the Zapatistas a lot: because things are so shit over here, something that looks brilliant, exciting, and a little bit different (perhaps there was a degree of exoticism in it as well), people overly invest in it and overly identify with it but of course it can't be transplanted wholesale to a different context.

So it's important to look at what's happening more locally too—rather than depoliticizing hope by displacing it onto an other—and thinking about where the connections might be. We've got anti-fracking campaigns, migrant solidarity campaigns, and certainly with the anti-fracking campaigns I think the political content of them is yet to be determined. A lot of it is NIMBYism, a lot (though not all) of it is middle class [and white], but that's what we've got. People don't come into struggle with perfect positions. People get involved in struggle because something is affecting them [or something they care about] and through contact with a whole host of people—activists, other people struggling, people reading texts—their political positions can change. Green and Black Cross are doing some really important work in anti-fracking struggles, sending observers to villages in Sussex that perhaps haven't seen a lot of political struggles or protest previously.[21] Of course, not all struggle will take the direction we want it to, but I think it's really important that we don't give up on it as inherently flawed from the beginning because then it will be captured by the Kingsnorths. The [fascist] British National Party made great play of localist environmental policy and you could easily see the far right jumping on anti-fracking campaigns.

21. Green and Black Cross are a mutual-aid organization providing legal support for environmental and social protest in the UK.

A: To add to this, it was very inspiring to see Black Lives Matter UK shutting down London City Airport, and talking about breathability and atmosphere. That's hugely linked to any environmental discussion of climate change in terms of pollution but also the simple fact that London is rapidly becoming unbreathable. What was brilliant about the BLM statement that came out was that they insisted that breathability is differentiated—that the problem with expanding London City is not that it affects the whole of London, but rather that it disproportionately affects the poor Black communities in Newham, where the airport is located.

Something I was excited about was the opening of a discussion around atmosphere and breathability, which would bring in the environment as a space where effects are differentiated. So that was an exciting moment, which I hope hasn't stalled because no one else took it up. It seems like the environmental movement missed that, and it's interesting that it has done very little about atmosphere and pollution in London. For me, that seems like a really axiomatic struggle that could be acted on immediately, and would massively improve the welfare and livelihood of systematically oppressed peoples.

So, I think it's very possible to already envisage what some kind of environmental activism in the UK might look like—it might not be as simple as targeting resource extraction, campaigns around pollution would be just as valid. In terms of displacement by flooding, that's something we are going to perhaps see more of, but pollution is something that's happening immediately. I would say that I remain deeply hopeful because people are making these moves towards realizing that the environment is a context rather than some kind of sole cause and, as environment is contextualized, I think we begin to see something quite hopeful here.

I don't see it as *a* movement, but as a series of deeply fragmented local insurgencies. That's what movements have always been. If you read Aldon Morris, who's a great sociologist of the Civil Rights Movement, he says it wasn't a movement but a series of local insurgencies which came to be seen as a movement because they acquired a force great enough that it was impossible to resist them.[22] I don't think we can model what we do now on the Civil Rights Movement, but it's important to remember that event, the archetypal movement, wasn't a movement. So, on this basis, in thinking about anti-fracking campaigns, all of them have the capacity to become very successful local insurgencies in which the demand ceases to be just about "we're going to stop this one thing" and becomes how how we can begin to act in solidarity with those whose lives are determined by catastrophe.

22. Aldon D. Morris, *The Origins of the Civil Rights Movement: Black Communities Organizing for Change* (New York: Free Press, 1986).

D: There's a great article by Aufheben written in 1994, "The politics of anti-road struggle and the struggles of anti-road politics." It outlines a lot of these issues in that movement: which sometimes was driven by NIMBYism, sometimes by environmental concerns, sometimes by moral concerns, sometimes by a more holistic Marxism.[23] What happens in those past movements, the historical memory, I think, is actually pretty important: in their struggles did they bring issues together to show how they were connected? How? That's of real use in determining how we organize against environmental destruction in the UK without the protofascist rhetoric of "Our England."

A: Yeah, and that refusal of "Spitfire Ecology," of Merrie England and green fields with an old fighter plane flying over them, is undoubtedly the danger. I think a refusal of the nationalist image of the land, as well as an embrace of the antinationalist possibility of a cyborg Earth—which simultaneously does not deny the possibility of an Indigenous nationhood—is the kind of contradiction we have to work through. This working through can't be didactic; it can't just be based in speaking, nor just in writing, nor can we just hope that if we fight hard enough it will all sort itself out in the end. I guess what I'm caught up in is some kind of social life where we practice speaking, writing, and fighting as if they had never been separate in the first place. That's why *BASE* makes me hopeful; it's a good place for some regenerative conversation, for some kind of lovingly antagonistic chatter.

23. This is a slight misremembering of the article, which can be found at libcom.org/library/m11-anti-road-aufheben.

REFUGES AND DEATH-WORLDS

First published November 2016

When we first began writing as a collective, we made almost no mention of migration or borders. Such an omission is an indictment, both of our own thinking and practice as a collective, and of the thought and politics we were engaging with. Orthodox and radical environmentalism alike frequently neglect those amongst the most affected by the ecological crisis—the people who are displaced by it. It is clear that our early work reproduced this omission. "Refuges and Death-Worlds" marks our first real attempt to engage with migration and the politics of the border. There are some important points here, most notably a critique of the category of the "environmental refugee," and an insistence on seeing the border as a relation. However, on rereading this piece after the BASE Magazine interview (originally published the following year), there is a striking inattentiveness to race. While it critiqued a politics in which, as Primo Levi posits, "every stranger is an enemy" we failed to attend to the fact that not 'every stranger' is the same. This is particularly notable in the section on the Mediterranean, which makes no mention of the antiblackness that drives the war on migrants. In reading this today, it is important to keep this omission in mind.

In *If This is a Man*, a memoir describing his imprisonment in Auschwitz, Primo Levi writes that:

> Many people—many nations—can find themselves holding, more or less wittingly, that 'every stranger is an enemy.' For the most part this conviction lies deep down like some latent infection; it betrays itself only in random, disconnected acts, and does not lie at the base of a system of reason. But when this does come about, when the unspoken dogma becomes the major premise in a syllogism, then, at the end of the chain, there is the *Lager* [camp]. Here is the product of a conception of the world carried rigorously to its logical conclusion; so long as the conception subsists, the conclusion remains to threaten us. The story of the death camps should be understood by everyone as a sinister alarm-signal.[1]

As we write, this alarm signal grows increasingly shrill. White supremacist and antimigrant populisms across the world are drawing sustenance from that "latent infection," constructing a system of reason premised on the hatred of "strangers" within and beyond the nation. Movements and organizing principles such as Black Lives Matter, No Borders, and #AmINext—at the same time—draw attention to and resist the structural connections between acts of violence otherwise dismissed as random and disconnected.

As scholars of populism have argued at length, "the people" to whom populists appeal, and in whose name they speak, do not preexist such appeals and such speech. Populists construct their own people, and from them, draw their supposed legitimacy and popularity. Terrifyingly, we can easily see how successful reactionary populists have been in this regard. The infection Levi speaks of is spreading and proving fertile for fashioning a xenocidal people.[2] Meanwhile, "the people" who might enact a *counterpopulism* remain only a latency. We can imagine, but not fully point to, a constituent power formed from the ensemble of those active in migrants' struggles—including migrants themselves.[3] Such a counterforce, George Ciccariello-Maher argues, could "subject institutions permanently and ruthlessly to popular pressure from below, to the demands of this tenuous, variegated multiplicity that is *the people*."[4]

---

1. Primo Levi, *If This Is a Man/The Truce* (London: Abacus, 2003), 15. If we take Aimé Césaire's point that the Holocaust had its roots in colonial genocides, then we should not be surprised that non-Europeans are more readily treated as enemies.

2. By xenocide, we reference the deliberate extermination of foreign entities. This term has roots in the practice of the intentional eradication of foreign plant or animal species.

3. For more on the role of "the people" in populism, see "Climate Populism and the People's Climate March" in Section IV of this volume.

4. George Ciccariello-Maher, *Decolonizing Dialectics* (Durham: Duke University Press, 2017), 133.

In 2017 there were a record 68.5 million forcibly displaced people in the world.[5] That was roughly one in every 105 people, and this figure only includes refugees and those displaced internally by armed conflicts. This figure would rise if those moving due to poverty, economic exploitation, or "natural" disasters such as droughts, storms, and desertification were included. Displacement is usually multicausal and attributing any given movement of people to climate change is difficult. Although the UN finds climate is already a factor in eighty-seven percent of disasters, such delineations are impossible to untangle from their social, economic, and historical conditions.[6]

The US government recently allocated its first funds for internally environmentally displaced people after a decade of relentless community and activist pressure, providing $48 million to relocate the Indigenous Biloxi-Chitimacha-Choctaw community of Isle de Jean Charles in southeastern Louisiana. This money was ultimately rejected by the community, who saw it as a ploy to erode their sovereignty and hijack a plan they had developed for years.[7]

In 2016, researchers of the Max Planck Institute for Chemistry and the Cyprus Institute in Nicosia argued that by the end of the century, "the Middle East and North Africa could become so hot that human habitability is compromised. The goal of limiting global warming to less than two degrees Celsius, agreed at the recent UN climate summit in Paris, will not be sufficient to prevent this scenario."[8] The Middle East and North Africa are currently home to around 400 million people. Another study found that with the same two degrees of warming, desertification is likely to push north through Morocco and into southern Spain. While some cities could adapt to increasingly hostile desert conditions given sufficient resources, such displacements in other low-latitude regions could mean one in twenty-five people becoming environmentally displaced by the twenty-second century. Without significant political change, "sufficient resources" will simply not be available in these regions for any but the richest inhabitants. The Adaptation Fund established under the Kyoto Protocol to facilitate projects

5. UNHCR, "Global Trends: Forced Displacement in 2017" (Geneva: UNHCR, 2018), 2. Available at https://www.unhcr.org/5b27be547.pdf.

6. UNISDR, "Ten-Year Review Finds 87 Percent of Disasters Climate-Related," accessed April 25, 2019, https://www.unisdr.org/archive/42862.

7. The terms of the resettlement plan are very much contested. See Julie Dermansky, "Critics Say Louisiana 'Highjacked' Climate Resettlement Plan for Isle de Jean Charles Tribe," *DeSmogBlog* (blog), April 20, 2019, https://www.desmogblog.com/2019/04/20/critics-louisiana-highjacked-climate-resettlement-plan-isle-de-jean-charles-tribe.

8. Max Planck Society, "Climate-Exodus Expected in the Middle East and North Africa," *Phys.org*, May 2, 2016, https://phys.org/news/2016-05-climate-exodus-middle-east-north-africa.html.

in the Global South funded by the "Annex I" countries of the Global North has allocated only US $358 million to adaptation projects in 68 countries since 2010. Existing finance mechanisms have also been critiqued for simply extending the hegemony of the Global North.[9] For comparison, in 2014 the UK announced a £2.3 billion (US $2.9 billion) spend on flood defenses alone, also over six years. The Adaptation Fund is not the only source of funding for such projects in North Africa and the Middle East, of course, but it contributes a significant portion of spending on climate change adaptation.

If temperatures increase beyond 2°C warming the Sahara Desert effectively jumps the Mediterranean.[10] At higher temperatures still, beyond the 4°C forecast for 2100, the world will be confronted with what Mark Lynas calls "zones of uninhabitability": areas in which "large-scale, developed human society would no longer be sustainable." Accordingly, he states, "we perhaps need to start talking about zones of inhabitability: refuge."[11] For us, there is no longer a "perhaps."

Much of Europe lies at temperate latitudes, meaning it could constitute one such refuge. Yet, even before such a likelihood we are told that Europe is experiencing a "migrant crisis." Rising numbers of displaced people seeking entry are coming up against increasingly fervent antimigrant populism, which denies their personhood in the name of xenocidal people.[12] One result of this crisis was the scaling back of search-and-rescue operations in the Mediterranean in order to deter attempts to cross. The Italian-led Operation Mare Nostrum was cancelled in 2014 and replaced with the far less extensive EU-led Operation Triton. Such moves are explicitly designed to stymie a politics of refuge. The switch from Mare Nostrum to Operation Triton resulted in a predictable and intended increase in deaths at sea. Indeed, even with relatively small numbers of attempts to migrate to Europe we see awful numbers of victims. 32,000 dead or missing between January 2000 and

9. David Ciplet, J. Timmons Roberts, and Mizan Khan, "The Politics of International Climate Adaptation Funding: Justice and Divisions in the Greenhouse," *Global Environmental Politics* 13, no. 1 (2013): 49–68.

10. Joel Guiot and Wolfgang Cramer, "Climate change: The 2015 Paris Agreement Thresholds and Mediterranean Basin Ecosystems," *Science*, 354 (2016): 465-468.

11. Mark Lynas, *Six Degrees: Our Future on a Hotter Planet* (Washington, DC: National Geographic Books, 2008), 209.

12. Rising numbers remain modest in light of forecast climate migrations and when compared to other parts of the world. Only a small fraction of the world's refugees try to enter Europe. According to the UNHCR's end-of-2017 figures, the European continent "hosted" 2.6 million refugees (this figure does not include Turkey, which hosts 3.5 million alone); compared to 6.3 million in sub-Saharan Africa and 4.2 million in the Asia-Pacific region. It is a similar case in the United States. The Americas as a whole "host" just 644,200 refugees (a *decline* of six percent from 2016). [See UNHCR, *Global Trends*.] This does not imply benevolence on the part of states hosting greater numbers, only proximity to displaced populations. In many states there is no route to citizenship, so people displaced decades ago from Palestine or the Soviet invasion of Afghanistan remain 'refugees', as do their descendents.

January 2016. And these are not simply deaths. Migrants are being murdered by the EU's border regime. The "migrant crisis" is, in fact, a border crisis whose underlying spirit is a barely concealed racial revanchism which says that *every stranger is an enemy*.

In his 1974 essay "Lifeboat Ethics: the Case Against Helping the Poor," ecologist Garrett Hardin proposed the metaphor of rich nations as lifeboats with a small amount of spare capacity surrounded by more swimmers than could safely be accommodated in the boats, "begging for admission". With no criteria to choose who to admit to the nation-boat, Hardin suggests the inhabitants should admit no-one. Just as the boat would sink so, he argues, would the nation be destroyed by the "fast-reproducing poor" as they overwhelm the "slow-reproducing" rich.[13] Hardin called this "lifeboat ethics," and it provides a ready rationale for the wholesale murder of migrants as a morally imperative act of racial-national self-defense. Not only is this morally repugnant, the supposed ecological theory underpinning it is incorrect.[14] However, these arguments remain ideologically useful to those looking for environmentalist justifications for border violence in an era of mass displacement.

To avert a future of lifeboat states, a solid understanding of existing border regimes is needed. An excellent place to start is with the concept of "border imperialism" developed by activists in the No One Is Illegal (NOII) network and fleshed out in Harsha Walia's *Undoing Border Imperialism* (2013). "Border imperialism," Walia writes, "can be understood as creating and reproducing global mass displacements and the conditions necessary for legalized precarity of migrants, which are inscribed by the racialized and gendered violence of empire as well as capitalist segregation and differential segmentation of labor."[15] Displacement has typically come through economic shocks, IMF structural adjustment programs, wars, and as we are seeing now, climate change will increasingly become a factor along with other aspects of ecological crisis. For example, the mining of raw materials not only produces carbon dioxide in the process of mining, through the uses to which industry subsequently puts those materials it causes all manner of pollution in the colonized, poorer regions where this generally takes place.

On paper, if not in practice, refugees have a legal right to refuge. States have enacted border imperialism by resisting and delegitimizing the category "environmental refugee." Border imperialism is predicated on a distinction between worthy and unworthy migrants. If "environmental refugee" comes to be legitimized

---

13. Garrett Hardin, "Lifeboat Ethics: The Case Against Helping the Poor." *Psychology Today* 8 (1974): 38–43.
14. See "The Dangers of Reactionary Ecology" in Section II of this volume for an expanded critique of Hardin.
15. Walia, *Undoing Border Imperialism*, 75.

by policy changes, we must attend to how states might use the category to further entrench distinctions between "good" and "bad" migrants. Rather than clinging to the rearguard issue of the right to asylum, perhaps we should be orienting our struggles towards the all-embracing demands raised by migrants: "freedom of movement for all," "everyone deserves a safe home," and "no more wall[s]."[16]

The notion of border imperialism calls our attention to the fact that the border is not just about lines on a map, it is something much more pervasive: the immigration raid on the workplace; surveillance in universities; nationality checks for school-age children, healthcare seekers, renters, and passport checks at train stations and bus stops. It is the riot police marauding through migrant camps and the activities of the EU border agency Frontex, for example, which "increasingly polices the EU's borders by taking its bordering practices directly to the populations it deems to pose the greatest threat," such as interdiction off the West African coast.[17] The border *is a relation*. Bordering practices produce conditions for the exploitation of precarious labor and "death-worlds" for those racialized as not fully human, not deserving of life.

Two quotes serve to illustrate this argument. The first is from philosopher Achille Mbembe:

> I have put forward the notion of necropolitics and necropower to account for the various ways in which, in our contemporary world, weapons are deployed in the interest of maximum destruction of persons and the creation of death-worlds, new and unique forms of social existence in which vast populations are subjugated to conditions of life conferring upon them the status of living dead.[18]

The second is from Abu Jana, a Syrian migrant:

> Let me tell you something. Even if there was a [European] decision to drown the migrant boats, there will still be people going by boat because the individual considers himself dead already. Right now Syrians consider themselves dead. Maybe not physically, but psychologically and socially, [a Syrian] is a destroyed human being, he's reached the point of death. So I don't think that even if they decided to bomb migrant boats it would change people's decision to go.[19]

16. These slogans can be seen in Guy Smallman's photographs of a 2015 migrant-organized demonstration in Calais.
17. Nick Vaughan-Williams, *Border Politics: The Limits of Sovereign Power* (Edinburgh: Edinburgh University Press, 2009), 28.
18. Mbembe, "Necropolitics."
19. Quoted in Patrick Kingsley, "Passport, Lifejacket, Lemons: What Syrian Refugees Pack for the Crossing to Europe," *The Guardian*, September 4, 2015, http://www.theguardian.com/world/ng-interactive/2015/sep/04/syrian-refugees-pack-for-the-crossing-to-europe-crisis.

Levi's warning—that the "end of the chain" of the logic that renders these strangers, these walking dead, enemies—haunts us. Lifeboat ethics are readymade refrains to rationalize and naturalize these horrors, to beget lifeboat states and the death-worlds of their border regimes. The latent infection diagnosed by Levi demands antifascist inoculation.

INFRASTRUCTURE AGAINST BORDERS

First published December 2016

In this essay, originally published as the second part of "Refuges and Death-Worlds," we attempt to deepen our critical understanding of antimigrant populisms before trying to imagine a politics that could counter them. Looking back, both efforts yield some definite successes. The use of Mitropoulos' concept of oikonomia to understand the imbrication of race, family, and nation still feels generative, as do the reflections on kinship that come later in the essay. There is a sense of a renewed critical and imaginative energy in this piece, which comes perhaps from its emphasis on the question of not just what is wrong, but what should be done.

INTRODUCTION

"No Borders" struggles within "temperate latitudes" need to keep in mind the longer-term likelihood of large-scale environmental migration from regions rendered uninhabitable by climate change. It seems likely that existing antimigrant populisms will draw on the ideas of reactionary ecology to demand that "lifeboat states" deploy border violence against outsiders. Countering this danger involves contesting the increasing levels of surveillance that characterize so much contemporary border work, including that undertaken in schools, colleges and universities, hospitals and medical clinics, the housing market, and the legal system. Combatting the new wave of surveillance-heavy border work will therefore involve disrupting immigration raids and fascist organizing, breaking the collusion between the nation-state and landlords, and mobilizing against immigration detention. Myriad other forms of migrant solidarity actions are needed.

Much of this activity is, necessarily, immediate and reactive. Countering a street demo by the far right, for example, is crucial but the future of large-scale climate displacement means there must also be a longer-term project of building culture and infrastructure that inoculates against equating strangers with enemies. This essay explores the basis of popular antimigrant and white nationalist mobilization and makes some suggestions regarding what an antifascist, pro-migrant politics could involve.

THE RACE-FAMILY-NATION NEXUS

[N]o, the masses were not deceived, they desired fascism, and that is what has to be explained.... How does one explain that desire devotes itself to operations that are not failures of recognition, but rather perfectly reactionary unconscious investments?[1]

The specter of racialized refugees, described as "swarms" and "cockroaches" by prime ministers, presidents, and mass circulation newsmedia, has been both manufactured and exploited by resurgent far-right politics. These include Trump's keynote promise to build a wall and make Mexico pay for it, the fascistic atmosphere of Brexit Britain, Modi's India, and surging far-right parties in France and Hungary. Fears of fantasized—sexually predatory—refugees gain affective resonance within wider racialized and sexualized anxieties, embedded deep in white supremacist nation-states,

---

1. Gilles Deleuze and Félix Guattari, *Anti-Oedipus: Capitalism and Schizophrenia*, trans. Robert Hurley, Mark Seem, and Helen Lane (Minneapolis: University of Minnesota Press, 1983), 253.

which then stick to the figure of the migrant. A particularly lurid example of this racial-sexual fantasy was provided by the cover of a 2016 edition of Poland's conservative mass-market current affairs magazine *wSieci*. The cover depicts a white woman clothed in an EU flag beset by grasping Brown hands and arms from all sides, accompanied by a headline proclaiming "the Islamic rape of Europe."

Angela Mitropoulos identifies this nexus of race, nation, and sexuality as *oikonomia*—the law of the household.[2] The household is important as the site of reproduction for property relations through both inheritance and marriage contracts. It is also the site of the reproduction for the (racialized) nation through sexual reproduction. The normative (monoracial, heterosexual, nuclear) household, though increasingly rare, is the foundation of statist futurity. It is perhaps the key institution in (re)producing citizens loyal to property, nation, and race.[3]

The "emotional conflation between family, race and nation" Mitropoulos identifies is evident in use of the term "cuck," the in-vogue white nationalist insult for supposed race traitors. The term alludes to a racialized, psychosexual anxiety about miscegenation, with penetration of the nation by "rapefugees" imagined as hordes of Black and Brown sexual predators.[4] Neither Trump's well-documented sexual predation (seen by many of his supporters as not only acceptable but laudable white heterosexual virility) nor indifference to abuse scandals with predominantly white perpetrators represents a contradiction in this affective configuration. The emotional nexus of the racialized nation and sexual entitlement allows such people to see both as acts which protect women from racialized foreigners in order to better reproduce the white nation.

BORDER REGIMES, REFUGES, AND SOCIAL REPRODUCTION

Understanding the unconscious investments of populist antimigrant sentiments in *oikonomia* broadens our understanding of what the longer-term work of building a pro-migrant, antifascist, anti-border-violence culture involves, in sometimes unexpected ways. Struggles over matters such as reproductive liberation, sexual violence, and the ability for LGBTQ+ people to exist freely in public are also struggles

---

2. Angela Mitropoulos, *Contract and Contagion: From Biopolitics to Oikonomia* (Wivenhoe, New York, Port Watson: Minor Compositions, 2012).

3. Mitropoulos, *Contract and Contagion*, 160.

4. The sexualized nature of this anxiety is evident in the fact that the term "cuck" has been lifted from pornography. It echoes Nazi Germany's preoccupation with "Rhineland Bastards," children born to German mothers who were fathered by occupying French colonial (mainly West African) troops after the First World War.

over the emotional core of racialized border violence. They are not peripheral, distracting "culture wars" or "identity politics."

Efforts to deny women and trans people bodily autonomy share logics with those to control the movements of racialized others across borders. The logic of reproductive futurism reproduces race and nation through the "proper" family and body politic. The UK policy requiring landlords to check their tenants' passports (enforced via a £3,000 fine for landlords who do not comply), effectively makes many people homeless and further endangers those fleeing domestic violence. The British government has also repeatedly challenged the sexuality of asylum seekers, as in the case of Nigerian LGBTQ+ rights activist Aderonke Apata, who was threatened with deportation to ultra-conservative Nigeria after being accused of "faking gayness" to get asylum status. Mitropoulos is thus correct to argue that it is "impossible to separate gendered and racial violence" and this is precisely how "men being entitled to regard women ([who] they read as white like them) as their property has been an important compensatory element in the history and politics of class and race." The imperative of "proper" racial reproduction lives on in the form of widespread anxieties about private property and paternity (that is to say, biogenetic property), the racial ordering of feminine "availability," and the policing of women's sexual "promiscuity."[5]

It is no surprise that the most viciously pro-border-violence politicians and media are increasingly anxious about the erosion of heterosexuality and binary gender norms. Despite the resurgence of quasi-fascist politics, they really feel like they're losing. And in some important ways, they are. Even on the Left, we are seeing increasing calls to reconfigure organization around a narrowly imagined (and thoroughly heterosexual) "white working class."

Jasbir Puar cautions against understanding any apparent departure from oikonomic norms as inherently threatening to capital. With gay marriage in mind, she writes, "the capitalist reproductive economy (in conjunction with technology: in vitro, sperm banks, cloning, sex selection, genetic testing) no longer exclusively demands heteronormativity as an absolute; its simulation may do."[6] But while capital is happy with a simulation of the heterosexual family, queer struggles contest the reproduction of border-desiring subjects, those bearers of authoritarian values who bash queers as readily as they do migrants.

5. Mitropoulos, "On Borders/Race/Fascism/Labour/Precarity/Feminism/Etc.," *BASE* (blog), October 29, 2016, https://www.basepublication.org/?p=107.

6. Jasbir K. Puar, *Terrorist Assemblages: Homonationalism in Queer Times* (Durham: Duke University Press, 2007), 31.

Such struggles read the scope of No Borders politics as expansively addressing the problem of social reproduction. In doing so, they also produce the possibility of "collective control over the material conditions of our reproduction" and of "new forms of cooperation around this work outside of the logic of capital and the market."[7] Glasgow Council's buddy schemes for migrants and the Glasgow Unity Centre's solidarity work are instructive here, as approaches that could potentially be replicated:

> The asylum seekers were placed [by the council] in empty flats in long neglected high-rise estates. Neighbors appointed by the council to welcome the new families took the job seriously, bringing the new arrivals from Kosovo, Pakistan, the Democratic Republic of Congo, into their communities, holding parties, bringing families from across the world together. When families were told they would not be given asylum their Scottish neighbors refused to let the Home Office remove them from the UK. Immigration officials who arrived in the early hours for 'dawn raids' on families were met by enraged Glaswegians who refused to let the Home Office take their new friends away. The demonstrations became widespread and saw the end of the dawn raids. Many thousands of people who had been threatened with removal, including many families, were allowed to stay in Scotland.[8]

This kind of longer-term mutual-aid work creates social and material infrastructures, a kinship where reproductive labor traditionally assigned to the household is partially socialized. As Silvia Federici notes, "it is through the day-to-day activities by means of which we produce our existence, that we can develop our capacity to cooperate and not only resist our dehumanization but learn to reconstruct the world as a space of nurturing, creativity and care."[9] Here kinship is figured not through biological proximity nor geographical adjacency, rather it is expressed through a shared relation to the world such that whoever exists within it lives in an affective, affectable closeness, holding and supporting one another. Such a move disrupts the oikonomic operations outlined above. Such kinship thus encompasses all refusals of the reproductive-futurist nexus of race-family-nation.

Being able to regenerate ourselves from transnational and multiracial networks of mutual aid makes it more difficult for the racialized, sexualized anxieties that fuel border violence to take root. Importantly, *this tendency already exists*. Working-class people improvise their social reproduction under conditions where the "proper"

7. Silvia Federici, *Revolution at Point Zero: Housework, Reproduction, and Feminist Struggle* (Oakland and Brooklyn: PM Press, Common Notions, Autonomedia, 2012), 111.

8. Maryline Baumard, "Give Me Your Tired, Your Poor . . . the Europeans Embracing Migrants," *The Guardian*. August 3, 2015, https://www.theguardian.com/world/2015/aug/03/europeans-who-welcome-migrants.

9. Federici, *Revolution at Point Zero*, 3.

nuclear family, even when desirable, is often not economically viable. Here we might see the beginnings of a populism from below: a collective, bottom-up construction of "the people" who might respond to Primo Levi's alarm signal.[10]

In addition to the erosion of the patriarchal, binary gender norms central to the oikonomic nexus, recent struggles like the movements of the squares, Occupy, and Black Lives Matter have spread beyond national borders. You do not have to subscribe to breathless accounts of the hyper-networked obsolescence of the nation-state to realize that transnational solidarities are far from the only ones on offer. Nevertheless, transnational solidarities do emerge from the affordances of the communications infrastructure. That struggles can and do find resonance on a transnational scale is certainly encouraging for a No Borders politics adequate to the scale of our ecological crises.

ANTIFASCIST INFRASTRUCTURES

Following the widespread circulation of images of the body of Alan Kurdî, a three-year-old child who drowned in the Mediterranean while trying to reach the Greek island of Kos from Bodrum in Turkey, even far-right British tabloids and newspapers published sympathetic coverage.[11] Media representations do not change individual views per se, but they can create a particular "structure of feeling" that empowers particular groups. Psychologists Steve Reicher and Alex Haslam refer to this operation as "meta-representation," arguing in relation to coverage of the photograph of Kurdî's body that:

> It could be that no-one has actually changed their minds, but suddenly those who abhorred the demonization of migrant[s] have realized that they were not alone. This would fit with a fascinating literature which suggests that the media doesn't so much influence what people think, but what those people think others think (meta-representations rather than representations). But this still matters because it affects what we are prepared to do. Once we feel that we are not alone, that ours is part of a collective voice, we are much more willing to act in public.[12]

10. See "Refuges and Death-Worlds," above.

11. The intended final destination of Kurdî and his family was Canada, where they had family. His aunt, a Vancouver resident, had applied for refugee sponsorship but was denied. Abdullah subsequently blamed Canada for the tragedy, which also killed his wife and Alan's brother. This shows the importance both of understanding the global nature of a crisis often held to be "European" *and* of transnational migrant solidarity. [See, for example: The Canadian Press, "Drowned Syrian Migrant Boy's Father Says He Blames Canada for Tragedy," *The Globe and Mail*, September 10, 2015, https://www.theglobeandmail.com/news/national/drowned-syrian-boys-father-says-he-blames-canada-for-tragedy/article26313666/.]

12. Steve Reicher and Alex Haslam, "A 'migrant' is not a migrant by any other name," *The Psychologist*,

This suggests alternative media infrastructures and effective use of sympathetic extant media channels are important for enabling collective action. If the creeping fascism of our present seems inescapable this reflects a media monopoly on meta-representation, ensuring that pro-migrant people feel alone even when they're not.

Alternative media infrastructures and sympathetic channels will not be sufficient in and of themselves, of course. Even if they could help develop pro-migrant confidence to the point it could function as "the will of the people," we do not support any position in which professional politicians are tasked with enacting that will. Even if "the people" are far from the racist mass they're assumed to be by so many commentators on, and advocates for, contemporary racist populisms. Broader infrastructures of nonbiological kinship will be necessary to enact and empower inoculation against Levi's "latent infection." Our preference is for the bottom-up self-organization of such infrastructures, and a diverse range of actors including, but not limited to, mutual-aid organizations, "local" communities, activists, artists, architects, legal professionals, international networks, and universities. Considerable attention to, and experience with, engaging uneasy alliances will be of importance in the coming years. This is what we mean when we refer to "a politics of regenerative cyborgs."[13] It builds on the already widespread forms of mutual support people construct for themselves alongside or in place of nuclear families. Bonds of affinity, solidarity, and kin-making replace those of blood, property, and nation.[14]

AGAINST INEVITABILITY

If even a fraction of projected climate migration takes place, reactionary forces can be expected to ramp up border panic and demand more border violence. Such programs will be organized around appeals to lifeboat ethics, with lifeboat states imposing death-worlds on racialized outsiders. Understanding the emotional resonance of these calls in the race-family-nation nexus centered on the household shows how supposedly unrelated struggles around social reproduction are in fact critical in contesting the reproduction of unconscious racial investments.

As the climate shifts to expand the world's uninhabitable zones, the nation-state as a mode of social organization in the habitable zones will come under

September 11, 2015, https://thepsychologist.bps.org.uk/migrant-not-migrant-any-other-name.

13. Donna Haraway, "The Promises of Monsters: A Regenerative Politics for Inappropriate/d Others," in *Cultural Studies*, ed. Lawrence Grossberg, Cary Nelson, and Paula A Treichler (New York: Routledge, 1992), 295–337.

14. See "The Future is Kids' Stuff" in Section III of this book.

considerable pressure. Its defenders will not likely accept its obsolescence lightly, and indeed, the military are making climate change central to their planning.[15] Authoritarian lifeboat states and associated genocidal border violence are not inevitable outcomes. While antimigrant populists are busy denying the existence of climate change, No Borders networks can get a head start by building on existing tendencies towards generosity and mutual-aid infrastructures to undermine the predictable reactionary responses when the climate crisis becomes undeniable even to the far right.

15. See Suzanne Goldenberg, "Pentagon: Global Warming Will Change How U.S. Military Trains and Goes to War," *The Guardian*, October 13, 2014, sec. Environment, https://www.theguardian.com/environment/2014/oct/13/pentagon-global-warming-will-change-how-us-military-trains-and-goes-to-war; Todd Miller, *Storming the Wall: Climate Change, Migration, and Homeland Security* (San Francisco: City Lights Books, 2017).

A HOSTILE ENVIRONMENT

First published November 2017

Almost exactly a year after the publication of "Refuges and Death-Worlds" we returned to the question of the border in this essay. There is a shift between these pieces that is immediately obvious; where "Refuges" elides race, "A Hostile Environment" seeks to center it, foregrounding the racial and the colonial in its opening paragraph. While this commitment is not entirely consistent throughout the piece, there is a notable desire to pay closer attention to the workings of the racial that prefigures our work that has come since. This piece also develops some of the ideas first published in the BASE Magazine interview, particularly the use of "critical dystopia," and the insistence on its differentiation by race, class, gender, sexuality, and disability. The final paragraphs, reflecting on the fact that the "hostile environment" has to be actively maintained, make a strong argument that this dystopia is not inevitable.

STATE POWER, WHITE POWER

The ascent of Donald Trump to the White House, and the frequency with which organized white supremacists have reappeared in Western democracies, has made it impossible to ignore or trivialize contemporary white nationalism. The fact that Trump has hung a portrait of slave-owner and Indigenous genocidaire President Andrew Jackson in the Oval Office serves as a reminder that white nationalism has long been central to the American republic.[1] Indeed, Black and Indigenous movements have consistently highlighted the continuity of contemporary forms of institutionalized white supremacy with the imperial past of the United States.

In the era of Donald Trump's presidency, xenophobic border violence increasingly converges with climate disaster. As such, we feel it's become necessary to highlight the ways struggles against national borders and against the racial geographies of capitalism more broadly are of central importance to the fight against climate chaos.

Trump's signature campaign promise was to "Build the Wall." Years later, no one seems to know whether the wall will ever materialize—though as of 2019, seventy-six miles of fencing has been expanded or reinforced, including those sections through the Tohono O'odham Nation and Quitobaquito Springs. Nor is there any real expectation that the physical wall itself would make any meaningful impact on criminalized migration, much of which involves overstaying one's visa rather than clandestine border-crossing. The wall, more than a policy instrument, is both a material and symbolic pledge of allegiance to whiteness. The wall exists in tandem with White House-issued incitements of police brutality towards racialized minorities.[2]

Europe has been supposedly experiencing a "migrant crisis" resulting in a similar push to harden its borders.[3] Much of the current immigration to Europe has been driven by wars and state interventions in Libya and Syria. This crisis has not really been one of numbers. For example, the majority of displaced Syrians are internally displaced within Syria, and the next largest groups reside in neighboring countries like Turkey, Lebanon, and Jordan.[4] Relatively few come to Europe: fewer

---

1. On Andrew Jackson, see Roxanne Dunbar-Ortiz, *An Indigenous Peoples' History of the United States* (Boston: Beacon Press, 2014), 95–116.

2. Lincoln Anthony Blades, "Donald Trump Is Being Accused of Endorsing Police Brutality," *Teen Vogue*, July 30, 2017, https://www.teenvogue.com/story/donald-trump-nypd-speech-police-brutality.

3. Claudio Minca and Alexandra Rijke, "Walls! Walls! Walls!," *Society & Space* (blog), April 18, 2017, http://societyandspace.org/2017/04/18/walls-walls-walls/.

4. United Nations High Commissioner for Refugees, "UNHCR: Total Number of Syrian Refugees Exceeds Four Million for First Time," UNHCR, July 9, 2015, https://www.unhcr.org/news/press/2015/7/559d67d46/unhcr-total-number-syrian-refugees-exceeds-four-million-first-time.html.

than one million for the whole twenty-eight-member EU in 2017, and two-thirds of that figure is accounted for by Germany and Sweden.

As in North America, the crisis is one of borders and whiteness. Xenophobic political parties have made ground in Poland, Hungary, Austria, and France. While acting as UK Home Secretary, Theresa May sent vans around London emblazoned with the message "Go Home or Face Arrest."[5] As Prime Minister, she oversaw the implementation of "hostile environment policies"—such as mandated passport checks in schools, hospitals, and rented housing—to make everyday life harder for migrants.

Under the leadership of PM May, British citizens, all members of the "Windrush generation," were wrongly deported, detained, and denied medical care and benefits. "Windrush generation" is a term for West Indian immigrants who arrived in the UK in the late 1940s to address postwar labor shortages (the first ship that arrived in 1948 was called *Windrush Enterprise*). Importantly, Windrush families entered Britain legally, as workers enlisted from Commonwealth countries in the Carribean, but many have lived in Britain for many decades without official government documents testifying to their status. Immigration laws passed in 2014 and 2016, architected by May, are responsible for the abruptly illegal status of the Windrush generation families. The ensuing political scandals May has encountered as a result of her wrongful treatment of Windrush generation families are direct consequences of hostile policy she has implemented. Such consequences are unintended only inasmuch as they have caused political embarrassment for the government.[6]

After Brexit, many were indeed told to "go home" while walking the streets where they lived, amid a record spike in hate crimes against Black, Brown, and Eastern European people.[7] On the other side of the Atlantic, Trump's attacks on trans people and reproductive autonomy, alongside his supporters' "cuck" anxiety, have similarly sought to make reproductive duty synonymous with falsely biological notions of womanhood, while reviving long-standing American fears of miscegenation.

5. Patrick Wintour, "'Go Home' Billboard Vans Not a Success, Says Theresa May," *The Guardian*, October 22, 2013, sec. Politics, https://www.theguardian.com/politics/2013/oct/22/go-home-billboards-pulled.

6. Alan Travis, "Immigration Bill: Theresa May Defends Plans to Create 'Hostile Environment,'" *The Guardian*, October 10, 2013, sec. Politics, https://www.theguardian.com/politics/2013/oct/10/immigration-bill-theresa-may-hostile-environment.

7. "'Record Hate Crimes' after EU Referendum," *BBC News*, February 15, 2017, sec. UK, https://www.bbc.com/news/uk-38976087.

The Right grasps what Mitropoulos identifies as the nexus of race-family-nation as key to the reproduction of capitalist class power.[8] This is how Trump, who has the lowest approval ratings of any president, maintains majority support: at the time of writing fifty-four percent of white voters still endorse his presidency.[9] "Whiteness" is not a neutral demographic descriptor. It has always meant identification with the ruling class, today symbolized by a hereditary millionaire!

Mitropoulos designates the race-family-nation nexus what the ancient Greeks meant by *oikonomia*, the relations that constitute the "management of the household" and etymological root of both economics and ecology.[10] Today, the normative force of oikonomia is realized through a system of nation-state-centric "border imperialism," Harsha Walia's term for the displacement of populations by war and structural adjustment followed by their capture in networks of border violence.[11] Borders serve to sift moving populations into exploitable precarious labor. For example, Germany's relative openness to refugees is inseparable from its labor shortages. Such surplus populations, when they become inconvenient, are marked above all by their proximity to death. Mobile "death-worlds" have proliferated around the world in the past decades, in places like the deserts north of the Rio Grande, aboard overcrowded vessels on the Mediterranean, in European migrant camps besieged by riot cops and abuse-filled detention centers like Yarl's Wood. Death-worlds are the flip side of oikonomic management of nations.

THE WORLD AS CRITICAL DYSTOPIA

The dystopia seems to write itself—something the left sometimes seems too keen to celebrate. Yet as Loretta Lees notes, focusing on the dystopic risks erasure of collective struggle and the premature foreclosing of hope.[12] In light of this, we suggest reading our contemporary moment as what Tom Moylan and Rafaella Baccolini have called a "critical dystopia."[13] Critical dystopia attends to the very real racialized effects of ecological crises as they intersect and co-constitute class, gender, sexuality, and disability. And, in contrast to the fatalistic tendency that Lees identifies with

---

8. Mitropoulos, *Contract and Contagion*.

9. Gallup, "Presidential Job Approval Center," November 1–14, 2019, https://news.gallup.com/interactives/185273/r.aspx.

10. Mitropoulos, *Contract and Contagion*; also see "Infrastructure Against Borders" in Section I of this volume.

11. Walia, *Undoing Border Imperialism*.

12. Loretta Lees, "The Urban Injustices of New Labor's 'New Urban Renewal': The Case of the Aylesbury Estate in London," *Antipode* 46(4), 2014: 921–47.

13. Baccolini and Moylan, "Conclusion: Critical Dystopia and Possibilities."

so many evocations of dystopia, it rejects any air of inevitability. The critically dystopian society is constantly beset by moments and spaces of (potentially) utopian collective struggles that operate within that dystopia, act against it, and prefigure a world beyond it.

In our contemporary dystopia of border regimes, one particularly immediate form of struggle is the constant, collective refusal of borders by migrants themselves. As Ethemcan Turhan and Marco Armiero argue, "a truly radical perspective cannot but see the practice of trespassing borders as a revolutionary act per se, sabotaging the state's control, questioning authorities, and rejecting the legitimacy of laws and regulations which protect and facilitate the movement of commodities but not of people."[14] This sabotage and rejection of control does not occur only through "trespass" but through the resistance to bordering as it occurs in everyday life—the work of Migrants Organise against the hostile environment in the UK, and the *gilets noirs* [black vests] struggle in France, for example. Although only a minority of environmentally displaced people currently cross international borders, business-as-usual warming of 4°C or more could render significant areas of the world inhospitable to human habitation.[15] Struggles against border regimes thus form an important part of climate politics. Refusal threatens the reproduction of the border and calls into being a series of secondary struggles against it, as people act in solidarity with migrants attempting to reproduce themselves, creating new infrastructures of solidarity in the process.[16] While these can be witnessed the world over, they often take place at the heart of colonial power.

Infrastructural struggles appear at borders all over the world, including both sides of the Atlantic Ocean. In Greece, anarchists have self-organized a multicity network of squats and mutual-aid resources to provide housing and social services for some of the thousands of refugees trapped there by the hardening of the EU's internal national borders.[17] In France, olive farmer and immigration activist Cédric Herrou has become a folk hero for helping African migrants cross the border from Italy and providing them with shelter and food. Arrested multiple times, his convictions were recently overturned because they were found to be "humanitarian" and legal according to the French Constitution. Herrou, who sometimes

14. Ethemcan Turhan and Marco Armiero, "Cutting the Fence, Sabotaging the Border: Migration as a Revolutionary Practice," *Capitalism Nature Socialism* 28, no. 2 (2017): 3.

15. Lynas, *Six Degrees.*

16. Walia, *Undoing Border Imperialism*; Natasha King, *No Borders: The Politics of Immigration Control and Resistance* (London: Zed Books Ltd., 2016).

17. Niki Kitsantonis, "Anarchists Fill Services Void Left by Faltering Greek Governance," *The New York Times*, May 22, 2017, sec. World, https://www.nytimes.com/2017/05/22/world/europe/greece-athens-anarchy-austerity.html.

works with others to organize aid for migrants, has said: "Our role is to help people overcome danger, and the danger is this border."[18] As discussed earlier, under Conservative PM Theresa May's leadership, the UK Government issued policies designed to make the state a "hostile environment" for asylum seekers, an approach which has been met with resistance. For example, Docs not Cops brings healthcare workers and activists together to campaign about the NHS' extension of passport checks, information-sharing with immigration enforcement, and charges of unfair treatment for migrants. Against Borders for Children's campaigning has helped bring an end to the collection of nationality and country-of-birth data collection by schools across England.

In North America, resistance to border imperialism has upped the ante since the inauguration of Trump in 2017, seeking to instantiate a widening network of "sanctuary cities" whose everyday culture is one of sanctuary in the streets.[19] A strategy that accepts too quickly the accepted parameters of the state's legal definition of sanctuary could be criticized from some state-sceptic quarters, but it has enabled broad coalitions of actors to work together and provides footing for more radical struggles to take hold.[20] Indeed, resistance to Trump's #MuslimBan and his draconian immigration policies have developed links between antifascist and Indigenous organizing networks. Buses are blocked, knowledge is shared, support infrastructures are developed, and allies are trained in effective solidarity in the face of police raids, sabotaging efforts by the state to arrest, detain or deport people.

The group No More Deaths/No Más Muertes provides water, first-aid, and other humanitarian aid to migrants along the US-Mexico border. These actions, in the words of border activist Scott Warren (No More Deaths), are "commensurate to a better ethic of treating both land and life with respect in the border region."[21] In the Arizona desert, the group's provision of basic services to migrants has been a target of Customs and Border Patrol (CBP) Officers and federal prosecutors, who arrest and prosecute with vigor. In some instances, charges are given an environmental veneer. In March 2019, four No More Deaths volunteers were sentenced to fifteen months probation and fined $250 each for leaving food and water in Arizona's Cabeta Prieza

18. Angelique Chrisafis, "Farmer Given Suspended €3,000 Fine for Helping Migrants Enter France," *The Guardian*, February 10, 2017, sec. World news, https://www.theguardian.com/world/2017/feb/10/cedric-herrou-farmer-given-suspended-3000-fine-for-helping-migrants-enter-france.

19. Megan A. Carney et al., "Sanctuary Planet: A Global Sanctuary Movement for the Time of Trump," *Society & Space* (blog), 2017, http://societyandspace.org/2017/05/16/sanctuary-planet-a-global-sanctuary-movement-for-the-time-of-trump/.

20. Walia, *Undoing Border Imperialism*.

21. Scott Warren, "In Defense of Wilderness: Policing Public Borderlands," *South Atlantic Quarterly* 116, no. 4 (2017): 871.

National Wildlife Refuge. Caught in the act by a Fish and Wildlife Service Officer in 2017, the charges were, officially, all environmental: entering a refuge without a permit, abandoning personal property, and driving in a restricted area. In his ruling, Judge Bernardo Velasco said that these activists "erode[d] the national decision to maintain the Refuge in its pristine nature."[22] The intent behind the veneer is clear in the case of Warren, who was arrested hours after No More Deaths posted video evidence of CBP officers destroying water jugs left in the desert. Warren faced federal prosecution for two felony counts of "harboring and conspiracy" which carries a sentence of up to twenty years. What was his alleged "crime"? Providing two "lost illegal aliens" with directions and then "food, water, beds and clean clothes" for three days in 2018.[23] After a mistrial, Warren was acquitted of all charges in November 2019 despite the state of Arizona assigning one of its top prosecutors to the case. This minor victory was accompanied by a ruling that the provision of such humanitarian aid could be legally supported due to Warren's religious beliefs.[24]

Resistance to border imperialism is frequently linked to struggles dismissed as "identity politics." In San Francisco, a large group of feminists blockaded the headquarters of ICE (Immigration and Customs Enforcement) on International Women's Day 2017 in an action they described as a "Gender Strike." They explained that "ICE is a direct manifestation of the worst forms of oppression faced by the most vulnerable women, queer and trans folks. . . . Ours is a feminism that must destroy every patriarchal wall or border."[25] Those who are marginalized, oppressed, and exploited by the uneven dystopia we inhabit are at the forefront of challenging it.

THE MAINTENANCE OF A HOSTILE ENVIRONMENT

Migration need not lead to tension or conflict. If it did, creating a "hostile environment" wouldn't have to be state policy. Our dystopia is not inevitable and it does not make itself. Transnational working-class struggles must be grounded in solidarity work against border imperialism. The counterrevolutionary work

22. Justin Wise, "Women Sentenced to Probation after Leaving Food and Water for Migrants in Arizona Desert," *The Hill*, March 3, 2019, https://thehill.com/latino/432382-women-sentenced-to-probation-after-leaving-food-and-water-for-migrants-in-arizona.

23. César Cuauhtémoc García Hernández, "This Man Could Go to Jail for 20 Years for Giving Migrants Food and Water ," *The Guardian*, May 10, 2019, sec. Opinion, https://www.theguardian.com/commentisfree/2019/may/10/scott-daniel-warren-migrants-food-water.

24. Ryan Devereaux, "Humanitarian Volunteer Scott Warren Reflects on the Borderlands and Two Years of Government Persecution," *The Intercept* (blog), November 23, 2019, https://theintercept.com/2019/11/23/scott-warren-verdict-immigration-border/.

25. International Women's Strike USA, "San Francisco—Gender Strike! Bay Area," March 8, 2017, https://www.womenstrikeus.org/event/san-francisco-gender-strike-bay-area/.

the state puts in to shut down transversal solidarity movements—for example, among Indigenous peoples and migrants across North America—is key to the reproduction of the border.

We should stress, climate change does not produce conflict or displacement by itself, as if a direct causal line could be tied between environmental conditions and social results. Climate change is made violent only by way of the inequalities historically produced political-economic institutions, from food markets to neglected public housing to private "security" apparatuses. As Neil Smith argues, "there's no such thing as a natural disaster."[26] Even as the climate becomes more adverse to our sustaining ecologies—from the food we grow, to extremes of precipitation and drought—it is states and markets that make the environment hostile. It has long been established that famines are not so much characterized by a decline in available food, but by a collapse in rural incomes and the loss of commons.[27] For example, under market conditions, food follows the money out of the area hit by a drought. Meanwhile, fresh water, as a renewable transnational resource, becomes a source of conflict when states impose national borders on its flows.[28]

Alongside the struggles to block fracking and oil pipelines, coal power and airport expansion, there is a "homefront" to an insurgency against ecological crisis. This one is focused on social reproduction under hostile conditions, and works against the violence of the present while opening up cracks in which an alternative future world flickers. Here, we have sought to highlight the ways numerous struggles against the proliferation of borders and myriad acts of migrant solidarity worldwide are unacknowledged sites of environmental politics.

The persistence of the Trump administration's antimigrant policies reinforced by an authorization of widespread nationalism and violence should not result in a doubling down on the liberal multicultural nation simply because it is better than the racist one. Instead, we need an abolition of the oikonomic race-family-nation nexus. Migrant solidarity and abolition of border imperialism must be understood as central to climate justice. Green politics must also be hostile to whiteness.

26. Smith, "There's No Such Thing as a Natural Disaster."
27. Amartya Sen, *Poverty and Famines: An Essay on Entitlement and Deprivation* (Oxford: Oxford University Press, 1982).
28. Mark Zeitoun and Jeroen Warner, "Hydro-Hegemony—a Framework for Analysis of Trans-Boundary Water Conflicts," *Water Policy* 8, no. 5 (2006): 435–60.

II

NATURES

INTRODUCTION

CYBORG ECOLOGY

It is impossible to read—and difficult to write—about the ecological crisis without encountering "nature," a word harboring manifold overlapping political meanings. Two political approaches to nature are particularly apparent in contemporary politics, each of which has supposedly liberatory and clearly reactionary uses. The first is a Promethean approach, in which nature represents an outside wilderness that must be subordinated or tamed. With this approach, nature has an immutability and scarcity to be overcome and very often a colonial frontier which men, nation-states or firms construct and must grapple with to produce order and value. Pitting itself in opposition to the Promethean conquest of nature, the primary counterhegemonic nature is romantic. No less modern than the Promethean, the romantic approach describes a nature of purity, scarcity, and untamability in need of protection. This nature too can be colonial or racist. Its heteropatriarchical politics are as apparent and reprehensible as the Promethean approach, and its local scale of action can serve reactionary as much as revolutionary ends.[1] In sum, Promethean and romantic approaches are two sides of the same coin, mutually bound together in a hegemony of contradiction.

---

1. Val Plumwood, *Feminism and the Mastery of Nature* (London: Routledge, 1993).

Promethean approaches to nature are readily apparent in everyday life. This is the vision of natural resources taken by energy and mining firms, states, and agriculturalists. A new leftist defense of such an approach has come from the authors of "#ACCELERATE MANIFESTO for an Accelerationist Politics" (2013), who "declare that only a Promethean politics of maximal mastery over society and its environment is capable of either dealing with global problems or achieving victory over capital."[2] In 2016, the authors of "The Xenofeminist Manifesto: A Politics for Alienation" contend that "if nature is unjust, change nature," and position themselves as technomaterialist and antinaturalist.[3] Some aspects of the Promethean approach might seem appealing to the Left: if we use our intellect in common, could we not invent labor-saving machines and organizational practices that conserve maximal amounts of human freedom and ecological life? It isn't necessarily a bad idea, but such abstraction relies on a fundamental forgetting of the productive and reproductive labor required to sustain the supply chains of resource extraction central to contemporary technology.

Many on the Left negate such techno-optimist visions through an inversion of the moral economy of technology and nature, sometimes through the same texts. Noted ecological Marxist John Bellamy Foster argues against technomodernist readings of Marx, that anticapitalist approaches must accept ecological limits and defend a localist "sense of place" against globalizing capitalism.[4] In direct contrast to the xenofeminist (XF) position which takes alienation as foundational, he argues alienation is historically specific: "capitalism leads to a loss of connection with nature, fellow humans, and community."[5] In short, the problem is that we are alienated from nature by capital.

Such a position seems acceptable at face value. And yet, the same story of nostalgia and loss of purity is central to an equally modern romantic tradition fully consistent with reactionary ideology. Like Iyko Day's diagnosis of North American settler politics of landscape, we see such a position as a form of "romantic anticapitalism." Such aesthetics and politics frequently rely on reductive European accounts of Indigenous peoples as well as anti-Semitic or xenophobic demonizations

2. Alex Williams and Nick Srnicek, "#ACCELERATE MANIFESTO for an Accelerationist Politics," *Critical Legal Thinking* (blog), May 14, 2013, http://criticallegalthinking.com/2013/05/14/accelerate-manifesto-for-an-accelerationist-politics/.

3. Laboria Cuboniks, "Xenofeminism: A Politics for Alienation" (2016) available at https://www.laboriacuboniks.net/. Republished as *The Xenofeminist Manifesto: A Politics for Alienation* (London and New York: Verso Books, 2018), 93.

4. John Bellamy Foster, *Ecology Against Capitalism* (New York: Monthly Review Press, 2002).

5. Fred Magdoff and John Bellamy Foster, *What Every Environmentalist Needs to Know About Capitalism: A Citizen's Guide to Capitalism and the Environment* (New York: Monthly Review Press, 2011), 77.

of rootless abstract migrants.[6] It is too easy to fall into essentialist understandings of race, gender, labor, and nation that reproduce our dystopia. To say that romantics *can* fall into such reactionary articulations of nature does not mean that all *do*. Yet we frequently see such narratives at work within tendencies and orientations of the Green Left. Why might this be the case?

Some of the connections between visions of nature and reactionary thought are surprisingly easy to trace. In the mid-1960s, Garrett Hardin coined the ubiquitous environmental concept "the tragedy of the commons." This thought experiment posits that without strict controls, population growth will outstrip natural resources. Consequently, Hardin argued that to avoid such a crisis, we must limit immigration, increase privatization, and remove support for food aid. Shockingly, this reductive scientific standpoint and its reactionary conclusions have become environmentalist common sense. Hardin's white nationalist assumptions, smuggled into popular science essays through his scientific credibility as a microbiologist, equate nature and nation to rationalize coercive state violence aimed at reproductive autonomy and nonwhite immigration. His metaphor of the crowded lifeboat rationalizes triage as a necessity of survival. These ideas have had influence far beyond the far right: we have encountered attenuated forms of them in the Green Party of England and Wales and direct-action eco-activist circles. One cannot attend a lecture on the state of the global climate in the US or UK without an audience member handwringing about the dangers of overpopulation.

Indeed, without clear class and antiracist politics, reactionary ideas of nature easily circulate within the Green Left. Climate activist, novelist, and poet Paul Kingsnorth calls for an environmentalism grounded in what he takes to be a timeless relation to place: the nation. Kingsnorth understands the nation as organically rooted in the natural landscape, and thus in conflict with "rootless ideology of the fossil fuel age," which he locates emerging from metropolitan individuals who align themselves with immigrants. Such an understanding of "rootless globalists" draws on familiar anti-Semitic tropes which can only benefit the far right's call for more stringent border controls. Indeed, Kingsnorth's claims about a longstanding identity rooted in national place are wildly ahistorical. His position is completely unmoored from even a cursory understanding of the history of borders and nations, which would demonstrate their relative recency as European and Euro-American institutions, and their rare coincidence with the transition zones between ecologies. The facile comparison Kingsnorth frequently draws between European and

6. Day, *Alien Capital*.

Indigenous understandings of "nation" fails to recognize Indigenous internationalisms central to the fight against empire. In "Lies of the Land: Against and Beyond Paul Kingsnorth's *Völkisch* Environmentalism," we argue that instead of equating nature with pristine wilderness encroached upon by foreigners and urban elites, only a *relational* notion of place makes possible another environmentalism. This politics would not reify borders and nation-states, but rather build solidarity in order to abolish them.

Surely, however, we must accept that the planet's resources are increasingly scarce and this poses limits to the conditions for human and nonhuman flourishing, right? In "The Political Economy of Hunger," we explore such questions by dispelling the Malthusian myth that world hunger results from a scarcity of food, and by implication, that feeding a larger population would necessitate the production of *more* food. Engaging the work of development economist Amartya Sen, we debunk this trope and explore the political possibilities it forecloses.

Developing food systems that can sustain healthy human and nonhuman systems beyond capitalist maldistribution will require new technical and organizational forms and new politics. In "Contemporary Agriculture: Climate, Capital, and Cyborg Agroecology," we discuss the polarization of recent movements into an anticorporate, pro-organic, peasant-agriculture camp and a corporate, pro-GMO, capitalist agriculture camp. We propose Donna Haraway's "cyborg agroecology" as a way to think through questions of dispossession, biotechnology, centralization of capital, and agrarian commons. While opposed to any simple tech-fix, we see no reason to throw the technobaby out with the capitalist bathwater. It is our view that many of the concerns with GMOs have more to do with capitalist restructuring of agriculture than biotech per se. Under the aegis of synthetic biology, recent biotech has adopted some commons-based practices such as open source libraries of biological "parts." This has been partly a result of anti-GMO struggles' hostility to corporate power.[7] And while capital can thrive on open source principles, we stand by our call to refuse conflations of scientific technique with capitalism on the one hand, and environmentalism with protection of a nature imagined as "untampered" on the other. We hope that our critiques of ahistorical principles of "modernity" upheld by Prometheans dispels concerns and accusations of "ecomodernism."

The final three essays engage the ideas of some thinkers who have taken up the problem of understanding the simultaneity of capitalist exploitation and

---

7. Adam Rutherford, *Creation: How Science Is Reinventing Life Itself* (New York: Current, 2013).

production of nature. In "James O'Connor's Second Contradiction of Capitalism," we discuss an influential eco-Marxist account that brings "the conditions of production," including ecosystems, within the scope of Marxist analysis. "Murray Bookchin's Liberatory Technics" engages the ideas of the Marxist-turned-anarchist-turned-communalist. Bookchin was an early critic of agrochemical pollution who also advanced a post-scarcity politics open to liberating technologies. Bookchin's philosophy of technology offers us a clear alternative to either the Promethean domination of nature or the conservative localism of the romantics. "Organizing Nature in the Midst of Crisis" is a critical review of historian Jason W. Moore's *Capitalism in the Web of Life* (2015). While affinities might be drawn between our approach and Moore's, our "messy cyborg politics" results in a different response to ecological catastrophe.

Though we hope our critiques are generatively expansive, we acknowledge these three essays give only a limited sense of the possibilities of contemporary ecological praxis. They are, after all, the ideas of three white men from the United States. Another charge might be lodged against us at this point: does our "cyborg ecology" leave us unable to distinguish among different political responses to ecological crisis? This would likely be the position of the Marxist polemicist Andreas Malm, whose book, *The Progress of This Storm* (2018), attempts to produce a historical materialist theory for a warming world.

Malm does not believe that any communist or revolutionary politics capable of explaining and confronting the climate crisis can be found in Neil Smith's "production of nature" thesis nor Haraway's "cyborg manifesto." He argues that any reference to the intertwined complexity of sociotechnological and ecological systems is simply postmodern "dissolutionism." Without a firm binary between nature and humanity, Malm fears, it is all mud. His political argument is that without an easily definable understanding of nature (e.g., as separate from humanity) it is not possible to develop a political program adequate to confronting the climate crisis. We cannot, Malm argues, distinguish between better or worse accounts of political struggle.

There is much we agree with in *The Progress of This Storm*. Malm argues that the tinkering of networks and signifiers is an apolitical dead end and urges us towards a revolutionary materialist approach to ecological catastrophe. Malm grasps the unevenness and depth of the ecological crisis within which we find ourselves. Yet we contest his argument that to posit nature as "constructed" or "produced" works against a revolutionary project, feeding into the moorless project of capitalism. Developing a more complicated and complex (and indeed, historically situated) account of nature and society does not in and of itself preclude ecological class struggle. This book, we hope, serves as some indication of that. If, as part of the climate crisis, "immigrants and other others can be framed as *external* enemies,"[8] it is in fact crucial to examine how concepts and constructions of nature have helped accomplish that externalization. To us, it is necessary to meaningfully engage with antiracist and anti-imperialist approaches to this multifaceted crisis.

8. Andreas Malm, *The Progress of This Storm: Nature and Society in a Warming World* (London and New York: Verso Books, 2018), 221. Despite our differences with Malm's views, we highly recommend his *Fossil Capital: The Rise of Steam Power and the Roots of Global Warming* (2016), which is a magisterial and horizon-expanding history on the origins of the so-called "Anthropocene." For Malm, as for us, ecological crisis is bound up not with innate human drive, nor with population growth, nor with a decline in wood, nor with industry per se, but with capitalism. Like us, he thus prefers the term "Capitalocene" (and is credited by Moore with coining it).

THE DANGERS OF REACTIONARY ECOLOGY

First published June 2014

Ecological politics in the Euro-American understanding is perpetually haunted by Malthusian themes—the assumption that population growth will outstrip natural resources, causing a return to a Hobbesian/ Mad Max war of all against all. In this early essay, we provide a primer on one of the most enduring myths used to uphold this vision of human nature: Garrett Hardin's "Tragedy of the Commons." In the time that's passed since the original writing of this essay, it has become apparent that our critique fell into Hardin's trap. Reading some of his notionally scientific work and observing reactionary assumptions, we equivocated whether Hardin could be considered an (eco)fascist thinker. In fact, we later learned Hardin had a parallel career as a far-right nativist and anti-immigration activist, deliberately smuggling his politics into his popular 'scientific' writing.

Hardin seems apposite with a whole range of columnists and academics who launder far-right views in popular outlets, using dog-whistle terminology and faux-intellectual arguments while also fraternizing with or writing for explicitly far-right organizations.

But our reliance—and the Left's more generally—on the demographic transition model is equally suspect. As numerous feminists have demonstrated, counterposing Hardin's mythos with a supposed universal social logos equally frames population change instrumentally, reducing gendered reproduction and bodily flourishing to data trends, especially of those in the 'experimental colonies' of the Global South, and frequently with an explicit anticommunist politics.[1] A more complex and attentive ecological politics is needed.

---

1. Among many others, see: Betsy Hartmann, *Reproductive Rights and Wrongs*, Revised edition (Boston: South End Press, 1999); Dorothy E. Roberts, *Killing the Black Body: Race, Reproduction, and the Meaning of Liberty* (New York: Vintage Books, 1999); Laura Briggs, *Reproducing Empire: Race, Sex, Science, and U.S. Imperialism in Puerto Rico* (Berkeley: University of California Press, 2003); Michelle Murphy, *The Economization of Life* (Durham: Duke University Press, 2017).

THE TRAGEDY OF CAPITAL

This essay functions as a short critical introduction to the oft-cited, though less often read, biologist Garrett Hardin. His most famous and influential concept is "the tragedy of the commons," a collective-action problem designed to show how sharing common resources without punitive laws would lead to inevitable ruin. Far from being descriptive of a problem, the tragedy of the commons is itself used to justify numerous actions for which the term "problem" would be a drastic understatement: for "stealing [I]ndigenous peoples' lands, privatizing health care and other social services, giving corporations 'tradable permits' to pollute the air and water, and much more."[1]

Hardin first set out the problem in a 1968 essay in the journal *Science*, but today the "tragedy of the commons" remains a staple of environmental ethics, ecological economics, and introductory climate science texts. Hardin described the thought experiment thus:

> The tragedy of the commons develops in this way. Picture a pasture open to all. It is to be expected that each herdsman will try to keep as many cattle as possible on the commons. . . . As a rational being, each herdsman seeks to maximize his gain. Explicitly or implicitly, more or less consciously, he asks, 'What is the utility to me of adding one more animal to my herd?' . . . the rational herdsman concludes that the only sensible course for him to pursue is to add another animal to his herd. And another; and another. . . . But this is the conclusion reached by each and every rational herdsman sharing a commons. Therein is the tragedy. Each man is locked into a system that compels him to increase his herd without limit—in a world that is limited. Ruin is the destination toward which all men rush, each pursuing his own best interest in a society that believes in the freedom of the commons. Freedom in a commons brings ruin to all.[2]

For Hardin, a degree of "coercion" is necessary to avoid this mutually assured ruin. The commons should be enclosed and sold as private property or managed by the state (with allocation "on the basis of wealth"; or by "auction," "merit, as defined by some agreed-upon standards," "lottery" or "first-come, first-served").

There are two key flaws in Hardin's assertions. First, despite being published in *Science*, Hardin doesn't present much actual evidence. It is only a thought experiment. Taking a corrective approach, in 1990 political scientist Elinor Ostrom

---

1. Ian Angus, "The Myth of the Tragedy of the Commons," *Monthly Review Online,* August 25, 2008. Available at https://mronline.org/2008/08/25/the-myth-of-the-tragedy-of-the-commons/. Our critique of the tragedy of the commons owes a great deal to Angus.
2. Garrett Hardin, "The Tragedy of the Commons," *Science*, Vol. 162, Issue 3859 (December 1968): 1243–48. Available at https://science.sciencemag.org/content/162/3859/1243.

published *Governing the Commons: The Evolution of Institutions for Collective Action*, a rigorous theoretical and empirical study of commons management systems around the world, such as fisheries in Nova Scotia and Turkey, or irrigation agreements in the Philippines and California.[3] Ostrom's work shows that commons resource management do not tend to mutual ruination via self-interested exploitation, but are frequently goverened by complex social agreements and norms to prevent overuse. Hardin subsequently conceded his argument only applied to 'unmanaged commons' rather than the commons in general, but recedingly few situations would meet such conditions.

Second, Hardin's argument presupposes the very relations it posits as the cure.[4] Hardin assumes each herder seeks to keep as many cattle as possible. Herders are therefore not subsistence producers, producing for their own consumption, but are producing for others. Furthermore, each of them is doing so competitively: herders are producing commodities for the market.[5] These herders are each rational utility-maximizing agents, with no social bonds, norms or relations with one another despite sharing a pastoral commons. Finally, for there to be a market for an ever-larger number of cattle, others elsewhere must lack access to commons' from which they could provide themselves with cattle. In other words, Hardin's tragedy presupposes an isolated commons in a sea of private enclosures.

Hardin presupposes the historically specific relations of capitalism, relations which were, in fact, only established following the widespread enclosure and privatization of commons.[6] Hardin's tragedy would be better called the tragedy of capital, for it shows only how capitalist relations of competitive production, without limit, for the market, tend to undermine the conditions of production.[7]

Hardin's argument is historically false, theoretically circular, and empirically dubious, yet it continues to play an important ideological role. It is still taught uncritically in ecology classrooms and cited by those who advocate privatization,

---

3. Elinor Ostrom, *Governing the Commons: The Evolution of Institutions for Collective Action* (Cambridge: Cambridge University Press, 1990). World Bank economists were questioning Hardin's orthodox view of the commons even before Ostrom's book appeared, but World Bank policy remained steadfastly committed to privatization and enclosure.

4. "To judge from the critical literature, the weightiest mistake in my synthesizing paper was the omission of the modifying adjective 'unmanaged.'" Garrett Hardin, "Extensions of 'the Tragedy of the Commons,'" *Science* 280, no. 5364 (1998): 682–83. Available at https://science.sciencemag.org/content/280/5364/682.

5. The emphasis Hardin puts on rational utility-maximization strongly suggests they're producing cattle as commodities for sale in the market.

6. Hardin's acceptance of state enclosure in "Tragedy" and of "socialism" in "Extension" as possible solutions does not change this, because state socialism and public ownership do not eradicate the production of commodities for exchange on the market.

7. This is what James O'Connor calls "the second contradiction of capitalism." See "James O'Connor's Second Contradiction of Capitalism" in Section II of this volume.

enclosure, and market competition as the solutions to the very problems caused by privatization, enclosure, and market competition. This is not to say that ecological crisis isn't a major collective-action problem within capitalist social relations, which sometimes plays out along the lines of tragedy. Rather, we suggest that the tragedy has its roots in historically specific social relations and not in a timeless rationality trap where "the alternative of the commons is too horrifying to contemplate."[8]

POPULATION IS NOT THE PROBLEM

Despite its influence on environmental economics, Hardin's primary problem throughout his work was population growth and his solution to this was eugenics. His 1968 essay argues that "the freedom to breed is intolerable" and asked "how shall we deal with the family, the religion, the race, or the class (or indeed any distinguishable and cohesive group) that adopts overbreeding as a policy to secure its own aggrandizement?" His answer was coercion: "a dirty word to most liberals now, but it need not forever be so. As with the four-letter words, its dirtiness can be cleansed away by exposure to the light, by saying it over and over without apology or embarrassment."[9]

To support this apparent necessity Hardin does not cite contemporary demography, but rather Thomas Malthus, the eighteenth-century moralist and reverend. Malthus claimed that population would grow exponentially while food production would only grow linearly. Without occasional catastrophe, hunger and misery would be permanent and insoluble features of human society, since population would always outstrip available resources for food.[10] Hardin had a PhD in microbiology. Population studies of bacteria are a core part of any microbiologist's training. Indeed, bacteria will reproduce exponentially, doubling in number each generation until their growth is checked by a limiting factor, such as exhaustion of nutrients.

Hardin seems to rely on Malthus' morality tale and his microbiologist's common sense without bothering to check whether human populations *actually* grow until checked by famine. In short, they do not. The countries where the population

8. Hardin, "The Tragedy of the Commons," 1247.

9. Hardin, "The Tragedy of the Commons."

10. Malthus is known today primarily for his population predictions which treated human misery as an unavoidable historical result. While predicting inevitable suffering in theory, he actively advocated for avoidable suffering by opposing social policy that would aid the poor and the advocation of measures that would require greater dependence on markets for subsistence. [See Michael Perelman, *The Invention of Capitalism: Classical Political Economy and the Secret History of Primitive Accumulation* (Durham: Duke University Press, 2000), 310–15.]

is stable or declining are not ones where there is famine, and countries where there are famines often have growing populations. Thinking about population in terms of self-enclosed "countries" not only ignores in- and out-migration, but also fails to take into account the entwined history of global markets and colonialism, which has been far more culpable for famine.[11] Furthermore, as Amartya Sen has shown, recent famines have not been caused by a lack of food but a lack of purchasing power to buy food.[12]

Human population within any given area is taken to be stable whenever the birth rate equals the death rate. If the stabilization of population is caused by famine, it would mean the death rate rises to match the birth rate. In fact, *both birth and death rates fall*. Before the advent of modern medicine, birth rates and death rates were high, towns were disease-ridden population sinks, and the population was therefore predominantly young and rural. Modern understanding of how diseases spread led to falling death rates, falling birth rates, urbanization, and an aging population.[13]

The seemingly exponential growth observed by Malthus was in fact the transition between the high birth/death equilibrium to the low birth/death equilibrium. It has been argued that this transition follows a pattern, generating a chain of positive feedback once it begins.[14] The supposedly "most-developed" countries began this transition several centuries ago and are now mostly at the higher, older, urban equilibrium. Population decline is even a concern in some places. Many so-called "less-developed" countries are not yet at the higher equilibrium, have younger, more rural populations, and are still experiencing rapid population growth. Regardless of these country-specific understandings of demographics (which are, it must be reiterated, incredibly complex and should be understood in relation to the historical specificity of coloniality and labor markets), UN demographers expect the world population to stabilize somewhere in the nine billion region.

---

11. Mike Davis, *Late Victorian Holocausts: El Niño Famines and the Making of the Third World* (London and New York: Verso Books, 2017).

12. Sen, *Poverty and Famines*.

13. Urbanization also reflects the forcible separation of the rural population from the land via enclosures and colonialism, but only really took off once urban mortality rates fell.

14. Tim Dyson, *Population and Development: The Demographic Transition* (London: Zed Books, 2013). Revisiting this essay, we look more critically upon the supposed universality of Dyson's demographic transition model. As Michelle Murphy argues, the demographic transition model essentially suggests different "societies" or "countries" are more or less "developed" based on a teleological, colonial, and Eurocentric model of time. The demographic transition model abstracts reproduction and fertility rates out of their historical context and ultimately authorized their management by scientists, economists, and states. [See Murphy, *The Economization of Life*.]

Writing in 1798, Malthus mistook the rapid growth phase of a sigmoid curve for an exponential one. In fact, Malthus' main goal was not to advance a theory of human ecology, but to make a political attack on the poor laws and the idea of raising workers' wages. In the 200 years since Malthus, much counter-evidence has accumulated and yet the same basic conservative assumptions remain.

LIFEBOAT ETHICS AND ANTHROPOCENE REACTION

The supposed problems of the tragedy of the commons and exponential population growth led Hardin to develop a highly influential moral theory in 1974 which he published in *Psychology Today* as "Lifeboat Ethics," with the provocative subtitle "the case against helping the poor."[15] The metaphor was chosen to counter the popular metaphor of "spaceship Earth" favored by more progressive ecologists. There is no world government, Hardin points out, and you can't have a spaceship without a captain (apparently). Rather, the planet should be understood as a collection of lifeboats, each of which represent a nation. These lifeboat nations, with their limited capacity, are threatened by desperate migrants who—upon entry—will displace the existing population as a result of their supposed tendency to higher rates of reproduction. The racist, patriarchal subtext is barely veiled.

Hardin purports to be a sober realist, the brave breaker of bad news: keep the immigrants out; sterilize profligate breeders; use famine to depopulate Africa (and keep the beaches pristine!). "Hey, don't shoot the messenger, just telling it how it is!" But it's a strange realist who assumes human society is analogous to bacteria without bothering to consult historical study—especially when the consequences of this analogy approach genocide. This is an early case of the familiar "capitalist realism" of "there is no alternative," which seeks to put its reactionary politics beyond question by invoking "reality," where reality is an evidence-free thought experiment.[16]

Hardin is often, and quite understandably, understood to be a fascist. After all, he invokes ecological limits to promote an eugenic agenda hostile to immigration and womens' bodily autonomy. Is this correct? Perhaps not: unlike fascists he does not call for a radical rebirth of the nation.[17] Rather, he seems to adopt

---

15. Garrett Hardin, "Lifeboat Ethics: The Case Against Helping the Poor," *Psychology Today* 8 (1974): 38. See 'Refuges and Death-Worlds', this volume, for a longer critique of lifeboat ethics.

16. Mark Fisher, *Capitalist Realism: Is There No Alternative?* (Winchester: Zero Books, 2009).

17. See Roger Griffin, *The Nature of Fascism* (London: Routledge, 1993). Griffin refers to this core aspect of fascism as "palingenetic ultranationalism."

the world-weary resignation of Cold War "realist" conservatism. And conservatism, of course, has a long history of anti-immigrant sentiment and eugenicist lust (Churchill as well as Hitler admired eugenics, nationalism, and empire). Hardin is content to survive on his lifeboat, so long as those in the water know their place. Fascists promise to raise the wreck, restore its former glory, and sail it again—if only the deadweight can be thrown overboard first.

But this considers fascism only at the macropolitical level of key state actors seeking the broad aggregate of a national rebirth. We agree with Gilles Deleuze and Félix Guattari when they argue that "what makes fascism dangerous is its molecular or micropolitical power, for it is a mass movement: a cancerous body rather than a totalitarian organism."[18] By this they mean the ways desires are produced, shaped, and circulate in much more localized ways, and here Hardin contributes much. His desire to cleanse the dirtiness of coercion by repeating it over and over, for example, finds contemporary expression in memes of "austerity nostalgia."[19] His lifeboat ethics, meanwhile, are hardly a fringe, far-right phenomenon—and are all the more dangerous for it.

In elaborating the ways micropolitics and macropolitics interface, Deleuze and Guattari write that "the motto of domestic policymakers might be: a macropolitics of society by and for a micropolitics of insecurity."[20] The population is bound to the state through the continuous production of insecurity and anxiety about terrorist Muslims, benefits scroungers, and immigrants swamping our precious national lifeboat. The bogeymen might be phantoms, but the felt insecurity that such imaginations produce is real. The recent gains for the far right in Europe have exploited this phenomenon even if they did not create it.[21]

---

18. Gilles Deleuze and Félix Guattari, *A Thousand Plateaus: Capitalism and Schizophrenia*, trans. Brian Massumi (Minneapolis: University of Minnesota Press, 1987), 215. Michael Rosen understands this when he writes: "Fascism arrives as your friend. It will restore your honor, make you feel proud, protect your house, give you a job, clean up the neighborhood, remind you of how great you once were, clear out the venal and the corrupt, remove anything you feel is unlike you." [Michael Rosen, "Fascism: I Sometimes Fear . . . ," *Michael Rosen* (blog), May 18, 2014, http://michaelrosenblog.blogspot.com/2014/05/fascism-i-sometimes-fear.html.]

19. At the time of writing, the UK was awash with variations on the already ubiquitous "Keep Calm and Carry On" poster which "represent . . . a very specific brand of contemporary ideology; that of austerity nostalgia [they] call on the viewer to trust the judgment of government, submit to its authority (in your best interest)." [spitzenprodukte, "Viva Miuccia! Cursory Notes on the Political T-Shirt," *Libcom.org* (blog), June 18, 2014, http://libcom.org/blog/viva-miuccia-cursory-notes-political-t-shirt-18062014.] See also Owen Hatherley, "Lash Out and Cover Up," *Radical Philosophy* 157 (2009): 2–7.

20. Deleuze and Guattari, *A Thousand Plateaus*, 216.

21. For an analysis of how mainstream "security" discourse fuels the far right, see Arun Kundnani, "Blind Spot? Security Narratives and Far-Right Violence in Europe," *International Centre for Counter-Terrorism (ICCT) Research Paper* (The Hague: The International Centre for Counter-Terrorism, 2012). Available at https://icct.nl/publication/blind-spot-security-narratives-and-far-right-violence-in-europe/.

The reactionary right-wing United Kingdom Independence Party (UKIP) have thus far adopted the idiom "nail colours to the mast" (referring to a determination never to surrender) to climate change denialism (and the fossil fuel finance this attracts). This may not be such a bad thing: do we really want to see UKIP's populism merge with Hardin's reactionary ecology? The specious notion of "Blitz spirit" demonstrates how disaster communities can be coded as nationally specific—indeed, as a rebirth of "what makes Britain great."[22] Lifeboat ethics and austerity nostalgia are already a toxic mix which thoroughly saturates official politics. Hardin's dismal ecology forms the first draft of Anthropocene-era reactionary politics. As the climate continues to deteriorate, we will likely see revisions of this reactionary politics across the political spectrum.[23]

22. Of course, what actually makes Britain "Great" is that it's larger than "*petit Bretagne*" (Brittany) across the water, a fact of geography that surely brings a lump to patriotic throats!

23. Angela Mitropoulos, "Lifeboat Capitalism, Catastrophism, Borders," *Dispatches Journal*, Issue #001 (November 19, 2018), http://dispatchesjournal.org/articles/162/.

LIES OF THE LAND

AGAINST AND BEYOND
VÖLKISCH ENVIRONMENTALISM

First published March 2017

It is crucial to think not only about the contemporary set of political relationships, but also possible future alliances and recompositions that might be germinating. One of the more insidious possibilities is a stronger alliance between the insurgent far-right extremism around the world and a nationalist pseudo-environmentalism. Though many contemporary far-right governments appear hell bent on enacting an almost accelerationist planetary death drive, others espouse a defense of nature in the name of the nation.[1] Though Paul Kingsnorth, antiglobalization activist turned antiglobalist poet, is not among the far right, we believe his continued normative espousal of an ahistoric 'nation' as the basis for environmental defense is incredibly dangerous. Much more can be done to demonstrate the continuities of such a stance with 'liberal environmentalisms' and the administrative politics of state sovereignty in general.

---

1. Bernhard Forchtner, ed., *The Far Right and the Environment: Politics, Discourse and Communication* (Abingdon and New York: Routledge, 2019).

ENVIRONMENTALISM AND THE "NEW" RIGHT

On Saturday, March 17, 2017, *The Guardian* published a lengthy essay by writer, poet, and climate campaigner Paul Kingsnorth. His "The Lie of the Land: Does Environmentalism Have a Future in the Age of Trump?" considers nationalist environmentalism a suitable response to our current ecological and political conjuncture.[1] It has been widely shared on social media and attracted praise from—among others—*The Guardian*'s political commentator John Harris and Greenpeace Senior Political Advisor Rosie Rogers.[2] This horrifies us. It is, quite simply, a dangerous piece. Its argument and logics must be rejected by those seeking to think through an environmental politics appropriate to the current era of climate change.

Kingsnorth finds inspiration in those he calls "the new populists," like Steve Bannon and Marine Le Pen, outlining a program that leads to a fascist environmentalism. This is terrifying but it is not without precedent: environmentalist and ecological politics in the West too often tend towards reactionary views. For example, UK Greens have advocated for reducing migration to "protect our environment," US environmentalists thought the sterilization of women in the Global South was a proper response to reduction of the global population, green anarchists have defended transmisogyny in the name of the "natural," and states around the world have utilized violence against Indigenous populations to create and "protect" National Parks.[3]

Far-right National Front leader Marine Le Pen, meanwhile, drew on several environmental themes during her 2017 campaign for the French presidency.[4] For

---

1. Paul Kingsnorth, "The Lie of the Land: Does Environmentalism Have a Future in the Age of Trump?," *The Guardian*, March 18, 2017, sec. Books, http://www.theguardian.com/books/2017/mar/18/the-new-lie-of-the-land-what-future-for-environmentalism-in-the-age-of-trump.

2. In a tweet on the day of the article's online publication, Harris called it "the best, truest thing I've read in ages." Rogers replied by suggesting she book Kingsnorth for the "Left Field" stage at the Glastonbury Festival, which she curates. [See twitter.com/johnharris1969/status/843075823947186176.]

3. On support for reducing migration, see a letter by UK Green Party members to *The Guardian* in July 2013, which lambasts then-leader Natalie Bennett for criticizing the UK government's creation of a (self-described) "hostile environment" for migrants. [Guardian Staff, "Letters: Many Greens Worried by High Immigration," *The Guardian*, July 25, 2013, sec. Politics, https://www.theguardian.com/politics/2013/jul/25/greens-worried-high-immigration.] On support for sterilization, note that Paul Ehrlich was encouraged to write *The Population Bomb*—frequently cited in support of forced sterilization programs—by David Brower, then Executive Director of the US environmental lobby group the Sierra Club (Brower also gave the book its title). On environmental transphobia, see Be Scofield, "How Derrick Jensen's Deep Green Resistance Supports Transphobia," *Decolonizing Yoga* (blog), May 24, 2013, http://www.decolonizingyoga.com/how-derrick-jensens-deep-green-resistance-supports-transphobia/. On violence against Indigenous peoples, see the monitoring and campaigning work of Conservation Watch (conservation-watch.org).

4. Le Pen called for "a move towards a 'zero-carbon' economy in France as well as more organic agriculture and a 'revolution in eating locally.'" [Michael Stothard, "Marine Le Pen Uses Environmental Issue to Broaden Appeal," *Financial Times*, January 26, 2017, https://www.ft.com/content/613eeb24-e3fc-11e6-9645-c9357a75844a.]

a chilling historical precedent, consider the Nazis' reliance on the work of early geographers and ecologists such as Friedrich Ratzel to promote *lebensraum*, the "living space" held to be necessary for the flourishing of a "pure" nation. Kingsnorth situates himself in the legacy of the antiglobalization movement which, although largely left-wing, has sometimes repeated or overlapped with fascist ideas and imagery.[5] While we focus on the essay itself in what follows, Kingsnorth is no stranger to reactionary nationalism.[6]

Below, then, we outline our key areas of concern with Kingsnorth's argument and connect them to broader errors in the way he understands the world. Although he attempts to distinguish between a "benevolent green nationalism" and the obviously less benevolent policies of the far-right, we show that no such separation can be made.[7] Indeed, the key oppositions that structure his argument are precisely those that structure ecofascism. Rejecting these, we close this essay by pointing to the possibility of antifascist and decolonial ecological struggle.

PEOPLE, PLACE, AND NATIONALISM

Kingsnorth opens his essay with an admission that he voted for the United Kingdom to leave the European Union. He proclaims astonishment that his friends in the "leftish, green-tinged world" had not done similarly, wondering why those who come from "a tradition founded on localization, degrowth, bioregionalism and a fierce critique of industrial capitalism" would vote to remain part of the EU.

At this point, over 30,000 people have died as a result of EU borders since the turn of the millennium and the EU routinely subjects migrants to appalling conditions at the camps it runs.[8] Yet this is not the source of Kingsnorth's ire; indeed,

---

5. Raphael Schlembach, *Against Old Europe: Critical Theory and Alter-Globalization Movements* (Abingdon and New York: Routledge, 2016).

6. We have serious concerns about the Dark Mountain Project, which Kingsnorth cofounded and editorially directs with Dougald Hine. Vinay Gupta—who has "been around the Dark Mountain story since before it had a name," has spoken at its festivals, and has been published in two of its books—has openly stated that he would "seriously consider helping out" a "credible and basically human ecological fascism" [See twitter.com/TimCWrites/status/843597773714931716.] At one point in the essay under discussion here, Kingsnorth rhetorically asks if he is "a fascist," as if to suggest that any such accusation would be patently absurd. Our concern here is not whether Kingsnorth is "a fascist," but rather to show how much of his environmentalism resembles fascism.

7. In 2019, an Australian fascist murdered fifty Muslim people at two mosques in Christchurch, New Zealand. In his manifesto, the killer describes himself as an "eco-fascist" and adheres to a vision of what he calls "green nationalism" in the face of "globalism."

8. Reporting on the condition of child migrants at the now-closed reception center on Lesbos, Tzanetos Antypas, head of the humanitarian organization Praksis stated that "there were some [children], I'm not kidding, whose hair had turned white. When we moved them to an open camp they chose to remain listless in their tents. After so many months incarcerated in such overcrowded conditions, I was told they had forgotten

migrants are notable only by their absence from his essay. Kingsnorth favors the UK's dominant (and alarmingly right-wing) framing of the EU: it erodes borders in favor of free movement and this free movement erodes cultural differences. Accordingly, he positions Brexit as "the people," "fueled . . . by a sense of place and belonging" seeking to take back power from "rootless" "globalists." For him, this is the key political division of our current moment. Regardless of whether or not Brexit achieves these aims (spoiler: it won't), the vote "exhilarates" Kingsnorth. Astonishingly, so does the election of Donald Trump.

Appeals to "the people" are common in political discourse and are a central feature of populist politics. But as a political subject (and actor), "the people" never preexists such appeals. Rather, it is constructed through them, and acknowledging this can be an important step in constructing a politics to challenge the status quo.[9] Kingsnorth elides this and presents his "the people" as self-evident fact. They are grounded in and belong to a timeless "natural" environment which coincides with "the nation."

Kingsnorth's nation is a social formation with "traditions, distinctive cultures . . . religious strictures [and] social mores." It is the source of "color, beauty and distinctiveness" and fosters a "belonging and a meaning beyond money or argument." Such "belonging" is held to be particularly strong in "traditional" places. Kingsnorth references the Standing Rock Sioux as exemplary and makes a passing reference to the Zapatistas.

One might be tempted to read Kingsnorth charitably. Perhaps he is proposing a radical understanding of "nation" (and the concepts associated with it) in line with that offered by many Indigenous peoples. But no, Kingsnorth draws on the social psychologist Jonathan Haidt, who explicitly accepts the aforementioned Le Pen's distinction between "nationalists" and "globalists." It is such ostensibly even-handed analysis that, in Haidt's words, puts "reasonable concerns about the integrity of one's own community" on the same moral footing as "the obligation to welcome strangers" that we suggest is dangerous.[10] Kingsnorth may try to distance himself from "angry nationalism" and Trump (while expressing "exhilaration" at their surge to power), but this can only ever work as a halfhearted

how to walk." [Quoted in Helena Smith, "Forgotten Inside Greece's Notorious Camp for Child Refugees," *The Guardian*, September 10, 2016, sec. World news, https://www.theguardian.com/world/2016/sep/10/child-refugees-greece-camps.] On the number of deaths due to borders, see themigrantsfiles.com/.

9. See Jason Frank, *Constituent Moments: Enacting the People in Postrevolutionary America* (Durham: Duke University Press, 2009).

10. Jonathan Haidt, "When and Why Nationalism Beats Globalism," *Policy: A Journal of Public Policy and Ideas* 32, no. 3 (2016): 48.

disavowal, given his take on "the nation" coincides with that of the (colonial or imperial) European state.

Consider Kingsnorth's conflation of Indigeneity with the European and Euro-American nation-state. This is a key rhetorical device for the white supremacist right—think of calls to protect "Indigenous" Britons, for example.[11] It is particularly abhorrent given that so many nation-states exist because of their genocidal dispossession of *actually* Indigenous populations, by which we mean those whose identities and ways of life are dynamic relationships with the more-than-human ecologies of particular places.[12] Kingsnorth's nationalists, by contrast, appear from static places as if *ex nihilo*. Except there is no France without the subjugation of the Berber populations of North Africa or the Haitians in the Caribbean. There is no United States of America without the destruction of Turtle Island or the enslavement of Africans. The borders separating Norway, Sweden, Finland, and Russia have divided up the traditional lands of the Indigenous Sami population, preventing them from continuing their traditions of fishing, herding, hunting, and trading. The relational histories of national borders demonstrate their fundamental historical construction as effects of global power relations.

Where Indigenous nations and populations have been decimated by the brutal violence of colonialism, Kingsnorth's "nation" is threatened by a nefarious fantasy of "globalism" that promotes migration and dissolves borders and supports multiculturalism while "enthusing about breaking down gender identities."[13] Accordingly, "border walls and immigration laws" are held to be "evidence of a community asserting its values and choosing to whom to grant citizenship." As with fascism, this "cultural" politics is in fact a racial—and racist—politics.

Kingsnorth regurgitates the anti-Semitic trope of globalism as "rootless."[14] In this essay, he raises the specter of "violent Islamism" to add weight to his claims. Not once does he mention that Muslims inhabit many areas of the world most affected by, and vulnerable to, climate change. Nor does he mention the Islamophobia that

11. James Mackay and David Stirrup, "There Is No Such Thing as an 'Indigenous' Briton," *The Guardian*, December 20, 2010, sec. Opinion, https://www.theguardian.com/commentisfree/2010/dec/20/indigenous-britons-far-right.

12. Kim TallBear, "Genomic Articulations of Indigeneity," *Social Studies of Science* 43, no. 4 (2013): 509–33.

13. The implicit transphobia of this statement is not the only time Kingsnorth exhibits the sneering language of alt-right fascists: elsewhere in this column he takes a dig at those who have (supposedly) told him to "check his privilege."

14. Werner Bonefeld, "Antisemitism and the Power of Abstraction: From Political Economy to Critical Theory," *Antisemitism and the Constitution of Sociology*, ed. Marcel Stoetzler (Lincoln: University of Nebraska Press, 2014), 314–32. It is worth noting that even without the "rootless" appendage the very concepts "globalism" and "globalist" function as anti-Semitic dog whistles.

drives the EU's policy of leaving migrants to drown.[15] Although Kingsnorth is right to say that "Green spokespeople and activists rarely come from the classes of people who have been hit hardest by globalization," his reference to Standing Rock is as close as he comes to rectifying this. Kingsnorth is of course completely ignorant of the hundreds of Native Nations who gathered at Standing Rock, making it only the most recent in a long tradition of what Lakota historian Nick Estes describes as "Indigenous internationalism."[16] Despite his hostility to those who fly, Kingsnorth also makes no reference to the recent Black Lives Matter UK shutdown of London City Airport, undertaken to highlight the racist dimensions of climate change.[17]

NATIONAL NATURES

In fact, "climate change" is mentioned only twice in Kingsnorth's essay, each time in relation to forms of environmentalism he pits himself against. There is not a single mention of climate change's devastating impact on food production. Rather, his environmental concern is driven by a privileged romanticism that culminates in the nation-state: "wild" nature contributes to "[t]raditions, distinctive cultures, [and] national identities," with some of that "color, beauty and distinctiveness." Kingsnorth's reinforcement of "distinctiveness" to describe both nature and culture under threat from "globalism" reinforces the danger at hand here.

This nature is referred to as the "birthright" of a nation, and in a disturbingly *völkisch* turn-of-phrase, Kingsnorth states: "You want to protect and nurture your homeland—well, then, you'll want to nurture its forests and its streams too." This desire to wrap forests in the flag clears the way for what critical scholar of eco-fascism Peter Staudenmaier calls a "deadly connection between love of land and militant racist nationalism."[18]

15. For Muslim populations and climate change, see: Naser Haghamed, "The Muslim World Has to Take Climate Action," *Al-Jazeera*, November 4, 2016, https://www.aljazeera.com/indepth/opinion/2016/11/muslim-world-climate-action-161103101248390.html. For a historic overview of European Islamophobia (and, indeed, the necessity of Islamophobia for the construction of Europe), see Gil Anidjar, *The Jew, the Arab: A History of the Enemy* (Stanford: Stanford University Press, 2003). As its title suggests, this also charts the historic imbrication between Islamophobia and anti-Semitism.

16. Nick Estes, *Our History is the Future: Standing Rock versus the Dakota Access Pipeline, and the Long Tradition of Indigenous Resistance* (London and New York: Verso Books, 2019).

17. Alexandra Wanjiku Kelbert, "Climate Change is a Racist Crisis: That's Why Black Lives Matter Closed an Airport," *The Guardian*, September 6, 2016, sec. Opinion, https://www.theguardian.com/commentisfree/2016/sep/06/climate-change-racist-crisis-london-city-airport-black-lives-matter.

18. Janet Biehl and Peter Staudenmaier, *Ecofascism: Lessons from the German Experience* (AK Press: Edinburgh, 1995). Available at theanarchistlibrary.org/library/janet-biehl-and-peter-staudenmaier-ecofascism-lessons-from-the-german-experience.

Recalling Kingsnorth's dig at those who challenge gender identities, we would add that this "love" of the land is a deeply gendered, thoroughly heteronormative romance. As Lee Edelman writes, "Nature [is] the rhetorical effect of an effort to appropriate the 'natural' for the ends of the state." Edelman goes on to argue that such "a statist ideology" works precisely by "installing pro-procreative prejudice as the form through which desiring subjects assume a stake in a future that always pertains, in the end, to the state, not to them."[19]

For Kingsnorth, the reproduction of the nation-state is inseparable from the reproduction of its "nature." His writing falls back on the imagery of Mother Earth: pure, bountiful yet fragile, a set of ideal feminine characteristics which can then be imposed on gendered subjects. The idealized reproduction of nature can then be used to discipline human reproduction, which is itself the precondition of the nation-state—after all, what is a "birthright" without births? When Kingsnorth talks of the desire to "nurture your homeland," which we understand as the implicit operation of what Edelman calls "installing pro-procreative prejudice."[20] The word "nurture" has a rich subtext of gendered labor. The quiet assumption here is that the nuclear family will continue to function and women will continue to perform (unwaged) carework. The future of Kingsnorth's nation depends on this.

When faced with this all-enfolding reproductive duress we should remember that "what is at stake [is] not the ability to reproduce, but the capacity to regenerate, the terms of which are found in all sorts of registers beyond heteronormative reproduction."[21] With this, Jasbir K. Puar pushes us toward an anticolonial "cyborg earth," which rejects the colonial, heteropatriarchal values of bounty, purity, and fragility, instead posing the possibility of liberated life.

The relentless coloniality of Kingsnorth's thinking is expressed again in his chosen example of (supposedly) "benevolent green nationalism." He cites US President Teddy Roosevelt's creation of the US National Parks as proof that nationalism can choose to define itself by "protecting, not despoiling, its wild places." Yet the creation of National Parks saw the forced relocation of thousands of Indigenous people. The existence of park systems is possible only because of long and ongoing histories of genocide and dispossession.[22] Kingsnorth is absolutely right that Roosevelt saw in the act of protecting nature "America's identity." But this identity,

19. Lee Edelman, *No Future: Queer Theory and the Death Drive* (Durham: Duke University Press, 2004), 52.

20. Edelman, *No Future*, 53.

21. Puar, *Terrorist Assemblages*, 211.

22. Isaac Kantor, "Ethnic Cleansing and America's Creation of National Parks," *Public Land and Resources Law Review*, Volume 28 (2007): 41.

Roosevelt maintained in his multivolume *The Winning of the West*, was directly tied to its imperial project of conquering the "world's waste spaces" through a ruthless and violent frontier vigilantism.[23] By ignoring Roosevelt's blatant racism and colonialism, Kingsnorth even undermines his own argument. In lauding the Standing Rock Sioux, he suggests that Indigenous populations are exemplary close-to-nature "nations," yet they are an obstacle to the flourishing of nationalist nature—opposing, for example, the federal government's authority to govern the Black Hills National Forest, among other public lands.[24]

Other aspects of Kingsnorth's fusing of environmentalism and nationalism fall apart under scrutiny. While geological features are often used in the drawing of national borders such that they acquire an air of natural permanence, regional ecologies do not match up to national borders (think again of the division of Sami lands). The most important ecological changes in the contemporary world are driven by global forces that nation-states can do little to challenge. Climate change does not respect borders.

Those on the Left might at least find some solace in Kingsnorth's naming of "neoliberalism" as one "global" formation opposing environmentalism. He references the "carbon-heavy bourgeoisie" and the "bankers" who threw "the people of Greece, Spain and Ireland to the wolves." Yet his criticisms are moralizing rather than structural. There is no account of the bourgeoisie's role in colonialism. Nor of the fact that bankers act as they do because that is what capital demands of them.

In fact, as Kingnorth's essay illustrates, fascists, too, are completely at home making such critiques of capital. His use of the term "globalists" leaves the door wide open for *specifically* anti-Semitic critiques. Given capitalism's ability to continue functioning with and in fascist regimes, such weak anticapitalism (or anti-neoliberalism) is in fact useful for capitalism. As Theodor Adorno and Max Horkheimer noted, fascism "seeks to make the rebellion of suppressed nature against domination directly useful to domination."[25] Trump, for example, shows us that such words are as true as ever.

---

23. Quoted in Greg Grandin, *The End of the Myth: From the Frontier to the Border Wall in the Mind of America* (New York: Metropolitan Books, 2019), 120.

24. Jeffrey Ostler, *The Lakotas and the Black Hills: The Struggle for Sacred Ground* (New York: Penguin Books, 2011). This contradiction is central to settler colonialism. In settler colonialism's expansionist, extractive guise, Indigenous populations are treated as part of "nature," which acts as a resource for extraction, a limit to growth, and a sink for waste. In its romanticist, protectionist guise, Indigenous populations are positioned as a threat to "beautiful nature": they are too lacking in scientific knowledge to understand how to protect it. Kingsnorth veers between offering (problematic) support for Indigenous populations resisting the first of these modes and adopting the second mode himself. Our approach rejects both of these, and is closer to (and draws on) many Indigenous understandings of nature. On this, see Enrique Salmón, "Kincentric Ecology: Indigenous Perceptions of the Human-Nature Relationship," *Ecological Applications* 10, no. 5 (2000): 1327–32.

25. Quoted in Bonefeld, "Antisemitism and Abstraction," 326–27. To set this anti-Semitism in historical context it would be important to engage the influence of Martin Heidegger on strands of the environmentalist

Kingsnorth promises the future to those who can successfully harness a carefully curated vision of a national birthright. That they "will win the day" is "as iron a law as any human history can provide." Such a position must be rejected. History does not have "iron laws." History is produced through struggle.

In mentioning the Zapatistas and the Standing Rock Sioux resistance to the Dakota Access Pipeline, Kingsnorth seems to know this too, at least on some level. The struggles of Indigenous peoples across the world are not, in any sense, equivalent to the protofascist, völkisch environmentalism he espouses. Of course, they have much to offer those seeking to develop an ecological politics within, against and beyond our current crises. As Lakota historian Nick Estes has demonstrated, central among these lessons is that Indigenous struggles for liberation have always had an internationalist political character, clearly completely unbeknownst to Kingsnorth.[26] Of course, for many people subject to colonial violence, "the nation" is an organizing frame, but to conflate the way the term is utilized here with the nationalism of European colonial states is deeply disingenuous. When Frantz Fanon argued "national consciousness . . . is not nationalism," he does not mean it as "the closing of a door to communication," rather, it is "the only thing that will give us an international dimension."[27]

In contrast to Kingsnorth's static, essentialist understanding of "place," Indigenous concepts of place are dynamic and relational. "Place," "land," and "territory" function as ways of understanding the relationships between people, animals, minerals, and plants across different scales. It is their dynamism on social, political, geologic, and biological levels that gives them their very "sense of place." These relationships do not separate out human society from the natural world, as Kingsnorth does, but see them as inextricably interwoven.[28] Learning from such

---

movement and where his work and Kingsnorth's overlap. Heidegger drew heavily on the Greek concept of *autochthony*, which names the way in which people are rooted in the environment of a specific region. In his philosophical writings he opposed this to the "rootlessness" of "modernity." Heidegger was, of course, a member of the Nazi Party, and in his diaries this "rootless modernity" is figured as *Weltjudentum* [world Jewry]. For more on Heidegger, autochthony and Nazism, see Stephen l'Argent Hood, "Autochthony, Promised Land, and Exile: Athens and Jerusalem Revisited" (PhD Thesis, 2006), https://scholarship.rice.edu/handle/1911/18918. The idea of "rootless" Judaism also fueled anti-Semitism in the USSR.

26. Estes, *Our History is the Future*.

27. Fanon, *The Wretched of the Earth*, 179. The lines between decolonial nationalism and supremacist nationalism are not always clear cut. As Maia Ramnath argues, postcolonial states have "perpetuated the same kinds of oppression and exploitation carried out by colonial rule, but now in the name of the nation." [Maia Ramnath, *Decolonizing Anarchism: An Antiauthoritarian History of India's Liberation Struggle* (Oakland: AK Press, 2012), 5.] The postcolonial nation is not the same as the decolonial or decolonized nation, and Ramnath notes that it would be churlish for anarchists to reject the concept of nation out of hand given that it plays such an important role in so many struggles against colonialism and white supremacy.

28. See, for example, the following: Nathalie Kermoal and Isabel Altamirano-Jiménez, *Living on the Land: Indigenous Women's Understanding of Place* (Edmonton: Athabasca University Press, 2016); Glen Sean Coulthard, "Place

understandings and exploring the resonances with what we call "cyborg ecology"[29] is key if we are to prevent the worst excesses of climate change from taking hold.

Many Indigenous and colonized people see the places they inhabit as being destroyed not by the *opening* of borders but by the very *imposition* of colonial borders in the first place. Accordingly, they play an active role in the migrant solidarity movements that will be of continued importance in providing solutions to the dismantling of borders necessary in a warming (or any) climate. Harsha Walia describes one such situation:

> In 2010, when 492 Tamil refugees aboard the *MV Sun Sea* arrived on the shores of the West Coast [of Canada] and faced immediate incarceration, Indigenous elders opened the weekly demonstrations outside the jails by welcoming the refugees. As their contributions toward a national day of action to support the detained Tamil refugees, the Lhe Lin Liyin of the Wet'suwet'en nation hung a banner affirming, 'We welcome refugees.' And as part of this same national day of action, Pierre Beaulieu-Blais, an Indigenous Anishnabe member of NOII-Ottawa, declared, 'From one community of resistance to another, we welcome you. As people who have also lost our land and been displaced because of colonialism and racism, we say Open All the Borders! Status for All!'[30]

Concern with culture, place, and identity does *not* imply nationalism as Kingsnorth understands it. Neither can border violence be glossed as simply "a community asserting its values." Nor do Indigenous and colonized people necessarily feel threatened by the challenges to gender norms that Kingsnorth so sniffily frames as part of a globalist agenda. Indeed, Western gender (and sexual) norms are—like borders—often seen as colonial impositions that have done much to damage gender roles, identities, and sexualities that do not meet these norms.[31] Exploring the

Against Empire: Understanding Indigenous Anti-Colonialism," *Affinities: A Journal of Radical Theory, Culture, and Action* 4, no. 2 (2010): 79–83; Vanessa Watts, "Indigenous Place-Thought and Agency Amongst Humans and Non Humans (First Woman and Sky Woman Go On a European World Tour!)," *Decolonization: Indigeneity, Education & Society* 2, no. 1 (April 5, 2013), http://decolonization.org/index.php/des/article/view/19145.

29. On cyborg ecology, see "Introduction: Cyborg Ecology" and "Contemporary Agriculture: Climate, Capital, and Cyborg Agroecology" in Section II of this volume.

30. Walia, *Undoing Border Imperialism*, 136.

31. See, for example, the following: Ifi Amadiume, *Male Daughters, Female Husbands: Gender and Sex in an African Society* (London: Zed Books, 2015); Sandeep Bakshi, "Decoloniality, Queerness, and Giddha," in *Decolonizing Sexualities: Transnational Perspectives, Critical Interventions*, eds. Sandeep Bakshi, Suhraiya Jivraj, and Silvia Posocco (Oxford: CounterPress, 2016), 81–99; Tamasailau Sua'ali'i "Samoans and Gender: Some Reflections on Male, Female and Fa'afafine Gender Identities," in *Tangata O Te Moana Nui: The Evolving Identities of Pacific Peoples in Aotearoa/New Zealand*, eds. Cluny Macpherson, Paul Spoonley, and Melani Anae (Palmerston North: Dunmore Press, 2001), 160–80; Sujata Moorti, "A Queer Romance with the Hijra," *QED: A Journal in GLBTQ Worldmaking*, 3, no. 2 (2016): 18–34. Engagement with such accounts should not lead us to the understanding that Indigenous and colonized societies have "the answers" to misogyny, homophobia, and transphobia; nor that they are always-already inherently superior to Western views on gender (see Moorti on this in particular), but they certainly provide ample evidence for debunking Kingsnorth's ignorance.

resonances and tensions between such approaches and calls to "queer" ecological activism is of considerable importance.

Paul Kingsnorth does not seem to be an out-and-out fascist. But his völkisch environmentalism opens the door to revanchist, heteronormative, neocolonial, and white nationalist currents that have long existed in parts of Western green politics. The "other environmentalism" of the movements and approaches discussed above is also an already existing one. It doesn't prefigure the kind of static world that Kingsnorth seeks, but in its dynamism and struggles (including internally), prefigures the flux and complexity of an ecologically just world. It simultaneously exists locally—in the cracks and interstices wrestled or protected from capitalism, the state, and colonialism—as well as globally, in the internationalist spirit of solidarity that will be essential if we are to reject ecofascism. It creates "the people" not as a static avatar of racialized nationhood but as a dynamic, heterogenous collective seeking to build a new world.

THE POLITICAL ECONOMY OF HUNGER

First published November 2014

In this early essay, we address another central aspect of Malthusian common sense: the premise that population growth is particularly dangerous because it will cause famine. It does not take much scratching beneath the surface to find that it is not population changes but capitalist markets which prevent people from accessing enough food. In 2019, we are more wary of the Polanyian critique we outline below, instead convinced by the influence of Mitropoulos' queer Marxist analysis of Polanyi's conservatism.[1] In fact, the arguments in this essay would perhaps be better served with a more robust theorizing of Marx's "primitive accumulation" as a mechanism for mediating relationships between producers and land. We leave the piece largely unedited as evidence of the cul-de-sacs that sometimes form in developing ideas.

---

1. Mitropoulos, *Contract and Contagion*; Angela Mitropoulos, "Corbynomics, Moral Economy and Saving Capitalism," *S0metim3s* (blog), July 17, 2018, https://s0metim3s.com/2018/07/17/corbynomics/.

In our earlier essays on the relationship between climate, class-based society and food, we focused on historical analysis.[1] In this essay, we continue our investigation up to the present day, with informed speculation about the future of food production in the context of global warming and climate chaos. First, however, we want to ask a more basic question: why do people go hungry?

COMMON SENSE: ABSOLUTE SCARCITY?

The intuitive answer to the question "why do people go hungry?" is that people must lack food. "Chronic hunger" is typically explained by the Malthus-influenced argument that population growth perennially outstrips food production. "Acute hunger," such as famine, is typically explained by the absence of food.

The Malthusian argument which underpins Garrett Hardin's reactionary ecology is a simple one. Thomas Malthus (1776–1834) claimed that that population grows "geometrically" (by which he meant exponentially), whereas food production grows "arithmetically" (linearly). The population will always grow faster than the food supply, and therefore chronic hunger will be ever-present. In making this claim, Malthus was motivated by politics: particularly his opposition to the English Poor Laws, which provided welfare (and workfare, in the form of the workhouse) for the destitute.[2] He also just made it up. As Danny Dorling writes, "he was not just wrong because he lacked imagination; he also cheated. It is now known that he even made up the correlation he used to try to suggest causation."[3]

However, Malthus' argument continues to be cited as if it's self-evident in both everyday conversations and scholarly works.[4] "If it had not been Malthus," Dorling continues, "it would have been some other fool."[5] A similar assumption of absolute scarcity informs the Food Availability Decline (FAD) theory of

---

1. OOTW's "Climate, Class, and the Neolithic Revolution" (2014) focuses on the emergence of agriculture after the end of the last ice age around 10,000 years ago, while "Class Struggles, Climate Change, and the Origins of Modern Agriculture" (2014) explores precisely this in the context of the "little ice age" of 1550–1850. Neither essay is included in this volume, but they are available on libcom.org: https://libcom.org/blog/climate-class-neolithic-revolution-09062014 and http://libcom.org/blog/class-struggles-climate-change-origins-modern-agriculture-18082014.

2. Malthus' opposition to the Poor Laws played a significant role in their replacement in 1834, which decreed that relief could only come in the form of the workhouse. *The Times* referred to this as "the Starvation Act."

3. Danny Dorling, *Population 10 Billion* (London: Constable, 2013).

4. For instance, David Cleveland, professor of environmental studies at UC Santa Barbara, states bluntly that "over the longer term *Malthus was right*. His fundamental observation seems incontrovertible." [Emphasis in original.]. This quote comes from the otherwise fairly critical *Balancing on a Planet: The Future of Food and Agriculture* (Berkeley and Los Angeles: University of California Press, 2014), 23.

5. Dorling, *Population 10 Billion*, 111.

famines, which was debunked by economist Amartya Sen in his hugely influential 1982 book *Poverty and Famines.*

Sen took several major famines as his case studies and found the FAD's approach was unable to explain both *why* people went hungry and *who* went hungry. The Bengal Famine of 1943 claimed 1.5 million lives, yet food production was only marginally below the previous year and was in fact higher than other years which had not seen famine. The Ethiopian famines of 1972–74 also saw only single-digit declines in food production. This was far too small to account for the 50,000–200,000 deaths due to starvation. In the 1974 Bangladesh famine, food availability actually hit a four-year per-capita high. In the Sahelian famine that peaked in 1973, drought did lead to significant declines in food availability, but Sen argues this fact alone could not explain who went hungry and where.[6]

SEN'S ENTITLEMENT APPROACH

In response to these observations, Sen developed a new theory to explain famines in terms of "entitlements." In a monetary economy, money entitles the owner to commodities of equal price. A rise in food prices, a decline in income, or an exhaustion of savings could all lead to an "entitlement failure": insufficient money to buy food and potentially hunger or starvation. Money is not the only form that an entitlement might take. Sharecroppers or peasant farmers may be entitled to consume (a portion of) their own production without market mediation. Pastoral nomads might similarly possess food entitlements outside of the monetary economy, as many are recipients of food stamps or similar welfare measures. For Sen, "the income-centered view will be relevant in most circumstances in which famines have occurred."[7]

Sen does not deny that declining food availability can be a factor in increasing hunger. He only claims this is mediated by entitlements, that is, social relations. Indeed, Sen claims that "food being exported from famine-stricken areas may be a 'natural' characteristic of the market which respects entitlement rather than needs."[8] Mike Davis has sought to sharpen Sen's analysis, which at times downplays the role of environmental factors. Drawing on the work of Amarita Rangasami,

---

6. The majority of deaths in famines are not directly caused by a lack of food; in both hot and cold climates, the most prominent causes of death are communicable diseases transmitted through the air, water or pests, which flourish as starving people migrate and seek refuge and food in unsanitary camps.

7. Sen, *Poverty and Famines*, 155.

8. Sen, *Poverty and Famines*, 162.

Davis contends notes that "the great hungers have always been redistributive class struggles."[9] But these are only attempts by capitalists and colonizers to redistribute food away from the poor and the colonized to the wealthy and the colonial powers.

With the near-global spread of enclosure and colonization, a large and growing proportion of agricultural production is commodity production for the market. Food is produced to be sold. Commodity production is not motivated by the use to which commodities are put but rather the prices they can fetch. If the sale of biodiesel or beef returns a sufficiently high price, agricultural land used to feed the local population will be put to use for the feeding of cars, cows or cotton. To quote the opening lines of Sen's book, "Starvation is the characteristic of some people not having enough food to eat. It is not the characteristic of there not being enough food to eat."[10]

THE POLITICAL ECONOMY OF HUNGER

The fact that there's enough food to feed everyone has slowly been acknowledged amongst the ruling institutions. For instance the UN's Food and Agriculture Organization (FAO) has stated recently that "there is sufficient capacity in the world to produce enough food to feed everyone adequately; nevertheless, in spite of progress made over the last two decades, 805 million people still suffer from chronic hunger."[11]

In the almost forty years since *Poverty and Famines* was first published, Sen's stress on the important confluence of social relations with agriculture and food access has been replaced by a more technocratic approach, which sees the problem of global "food insecurity" simply as a matter of policy tweaks. "Food availability" remains the first term on the FAO's list of dimensions of hunger. And while between a third and a half of world food production is currently wasted, the World Bank, like Malthus, invokes a growing population to suggest that raising agricultural productivity must be a primary goal. There's nothing wrong in principle with increasing agricultural productivity; indeed, more output for less inputs seems like a good idea. But if productivity increases are to be achieved through price signaling alone, then land must be subject to the forces of the market as well. This can be achieved only through force—licit or illicit.

9. Davis, *Late Victorian Holocausts*, 22.

10. Sen, *Poverty and Famines*, 1.

11. Food and Agriculture Organization of the United Nations, "FAO's Strategic Objective 1: Help Eliminate Hunger, Food Insecurity and Malnutrition" (2015), http://www.fao.org/3/a-au829e.pdf.

Land grabbing is a form of primitive accumulation through which noncapital-ist modes of production are transformed into capitalist modes, what the Midnight Notes Collective refers to as "new enclosures."[12] These new enclosures dispossess and proletarianize rural populations, making them dependent on the market for subsistence. Following Marx, primitive accumulation is not only achieved through actual outright dispossession of land, but also through the provision of individual private property titles which international NGOs and global governance institu-tions have touted as solutions to both food and land insecurity—and, more fre-quently now, with reference to supporting *women's* land rights in particular.[13] Thus even while Sen's insights might be formally acknowledged by such institutions, policy emphasis quickly regresses to such rote capitalist approach of increasing output. To understand why this is, we turn to the economic historian Karl Polanyi.[14]

Polanyi was interested in "the great transformation": the rise of the modern market society. Like Karl Marx before him, he identified a three-stage separation of the population from the land as the key factor in the transformation of markets from fringe phenomena to the central institution governing social reproduction:

> The first stage was the commercialization of the soil, mobilizing the feudal revenue of the land. The second was the forcing up of the production of food and organic raw materials to serve the needs of a rapidly growing industrial population on a national scale. The third was the extension of such a system of surplus production to overseas and colonial territories. With this last step land and its produce were finally fitted into the scheme of a self-regulating world market.[15]

OOTW's chronology differs from Polanyi's. In ours, colonial production precedes and helps finance the Industrial Revolution. James Watt's steam engine, for ex-ample, was financed by profits from the West Indies slave plantations.[16] As such, hunger is not an incidental problem in capitalism but *a condition of its possibility.* More importantly for the matter at hand, Polanyi stresses that "the critical stage [in the early modern transition to capitalism] was reached with the establishment of a

12. Midnight Notes Collective, "The New Enclosures," in *Midnight Oil: Work, Energy, War, 1973–1992* (Brooklyn, NY: Autonomedia, 1992), 317–33. On "land grabbing" as a form of primitive accumulation, see also the follow-ing: Silvia Federici, "On Primitive Accumulation, Globalization and Reproduction," *Re-Enchanting the World: Feminism and the Politics of the Commons* (Oakland: PM Press, 2018); Glen Sean Coulthard, *Red Skin, White Masks: Rejecting the Colonial Politics of Recognition* (Minneapolis: University of Minnesota Press, 2014).

13. See, for example, Kelly Askew and Rie Odgaard, "Deeds and Misdeeds: Land Titling and Women's Rights in Tanzania," *New Left Review*, no. 118 (2019): 68–85.

14. See our introduction to Section II for a critique of our treatment of Polanyi in this essay.

15. Karl Polanyi, *The Great Transformation: The Political and Economic Origins of Our Time* (Boston: Beacon Press, 2001), 188.

16. Eric Williams, *Capitalism and Slavery* (Chapel Hill: University of North Carolina Press, 2014 [/1944]), 102.

labor market in England, in which workers were put under the threat of starvation if they failed to comply with the rules of wage labor. As soon as this drastic step was taken, the mechanism of the self-regulating market sprang into gear."[17]

This process of proletarianization created the category of the unemployed, which superseded that of the pauper. Polanyi argues that unless the unemployed were "in danger of famishing with only the abhorred workhouse for an alternative, the wage system would break down."[18] For this reason, Polanyi thought that the post-WWII welfare state and the Keynesian policy of full employment had, in minimizing the threat of hunger, superseded the market society. But social democracy and welfare turned out to be an unstable compromise between capital and organized labor, which collapsed as soon as new markets and labor pools could be found in the Global South. Following the economic crises of the seventies, capitalists responded with a renewed round of economic liberalism.

The return of rickets, food banks, and the workhouse (in the guise of workfare) should be seen as a return to capitalist normality. Capitalism needs to maintain this artificial scarcity of food to underwrite the labor market. Climate change is likely to damage crop yields and reduce available agricultural land through desertification, salination of coastal aquifers, and flooding from sea-level rises or changing precipitation patterns, not to mention poorly managed levees and concrete-covered cities. Food availability is always mediated by social relations. As Rolando Garcia puts it, "climatic facts are not facts in themselves; they assume importance only in relation to the restructuring of the environment within different systems of production."[19] Discussions of world hunger almost invariably assume that food production is and will continue to be commodity production, while simultaneously assuming that food is produced for use. But whatever the specifics of the conditions that climate change and broader ecological crisis will create, there is always a gap between what is possible and what is possible *in capitalism*. Declining crop yields and loss of arable land can be expected to increase world hunger. The social relations through which biophysical forces are organized are not 'laws of nature': they are subject to change. This is the revolutionary possibility that Malthusian mythology serves to obscure.

17. Polanyi, *The Great Transformation*, 225.

18. Polanyi, *The Great Transformation*, 232. Here we might remember Malthus' role in leaving the workhouse as the "only" alternative to famishing in Britain.

19. Quoted in Davis, *Late Victorian Holocausts*, 21.

CONTEMPORARY AGRICULTURE

CLIMATE, CAPITAL, AND CYBORG AGROECOLOGY

First published July 2015

Here, we continue our exploration of the roots of food insecurity in capitalist markets, while exploring the range of global social movements that press against the reduction of ecological systems to private property and value. While we reject outright the modernist idea that technological fixes will be able to avert ecological crisis, we are also wary of the tech-skeptical organicism, which can misplace the problem. Though today some of our analysis of the scientific and logistical underpinnings of contemporary agriculture might be more integrated with capitalism than we would hope, we maintain that any transition to communism will have to grapple with re-purposing an inextricable mix of so-called 'natural' and 'technological' forms: hence, cyborg agroecology.

The essay originally referred to this as 'cyborg ecology', but here we scale the name back as what we outline here pertains primarily to the agricultural rather than to the totality of ecology. A more holistic 'cyborg ecology' could be more explicitly theorized by piecing together from aspects of other essays in this book, as well as the practices and thinkers we engage with. This may well be a task we return to in future.

The essay originally referred to this as 'cyborg ecology', but here we scale the name back as what we outline here pertains primarily to the agricultural rather than to the totality of ecology. A more holistic 'cyborg ecology' could be more explicitly theorised by piecing together from aspects of other essays in this book, as well as the practices and thinkers we engage with. This may well be a task we return to in future.

Agriculture is hugely vulnerable to climate change, for obvious reasons. It is also the sector which is the basis for all other orders of change. In order to address this, UN bodies have focused on a "food security" approach, incorporating many controversial practices such as transgenic (genetically modified) crops, high inputs of agrochemicals and water, and the increased integration of farmers into commodity and financial markets. Against this model, agrarian social movements spearheaded by La Vía Campesina (LVC), a global organization which coordinates peasant movements and struggles representing over 200 million people worldwide, have proposed a program of "food sovereignty" based on localized production and distribution, and organic farming methods. While sympathetic to this latter approach, we find the strict distinction between traditional and modern methods closes off important possibilities. We propose to rethink these questions of social relations, nourishment, technology, anti-capitalist struggle, and scientific and practical knowledges through the lens of what we call "cyborg agroecology." This rejects dichotomies such as natural vs. unnatural, local vs. global, and fast vs. slow. Instead, we focus on the relationships (social, political, economic, more-than-human) determining agricultural production and distribution.

CLIMATE CHANGE AND FOOD SECURITY

In 2004, the Intergovernmental Panel on Climate Change's Fifth Assessment Report ("AR5") reported: "Negative impacts of climate trends have been more common than positive ones. . . . Since [2007], there have been several periods of rapid food and cereal price increases following climate extremes in key producing regions, indicating a sensitivity of current markets to climate extremes, among other factors. Several of these climate extremes were made more likely as the result of anthropogenic emissions." The report further finds that while some regions, mainly northern high latitudes, could see increased agricultural yields, on balance the impact on yields is likely to be negative. In the near-term, the impacts are not catastrophic, with only ten percent of projections showing yield losses of more than twenty-five percent, compared to the late twetieth-century levels. However, "after 2050, the risk of more severe impacts increases."[1]

---

1. Intergovernmental Panel on Climate Change (IPCC), *Impacts, Adaption, and Vulnerability, Part A: Global and Sectoral Aspects. Contribution of Working Group II to the Fifth Assessment Report of the Intergovernmental Panel on Climate Change*, 1st edition (New York: Cambridge University Press, 2014), 488.

Business-as-usual global warming puts average global temperatures on course for 4–6°C warming by 2100. The IPCC warn that "[g]lobal temperature increases of ~4°C or more above late-twentieth–century levels, combined with increasing food demand, would pose large risks to food security globally and regionally. . . . Risks to food security are generally greater in low latitude areas."[2] Parsing the IPCC's technocratic language from their previous reports into plain English, Mark Lynas writes that "it's difficult to avoid the conclusion that mass starvation will be a permanent danger for much of the human race in a four-degree world."[3]

As the above reference to food price rises and "low latitude areas" suggests, these impacts are unevenly distributed. Mass hunger is mediated by market dynamics and doesn't necessarily require an absolute scarcity of food. Indeed, today hundreds of millions go hungry, while at the same time up to half of the world's food supply goes to waste and substantial areas of land are dedicated to producing cattle feed and biofuels. A 2010 editorial in *Nature* noted: "[t]he 2008 food crisis, which pushed around 100 million people into hunger, was not so much a result of a food shortage as . . . market volatility."[4]

In extreme climate scenarios, the technical possibilities for feeding the world's population exceed the economic "optimum," which calculates not in terms of needs but ability to pay. As the editors of *Nature* note, "climate change adds a large degree of uncertainty to projections of agricultural output, but that just underlines the importance of monitoring and research to refine those predictions.

For officials at the Food and Agriculture Organization (FAO) of the United Nations, the practicability of feeding everybody, and indeed of producing the food required to do so, is premised on the idea "all the options are on the table," including controversial practices such as transgenic crops (GMOs), the "modernization" of land tenure (new enclosures), further integration of small farmers into financialized markets, and the use of synthetic fertilizers and insecticides. Practices such as these have become flashpoints in the polarization between two visions of world food production: one multinational biotech-led, the other the "peasant" alternative promoted by social movements like La Vía Campensina and ecofeminist activists like Vandana Shiva.

2. IPCC, *Climate Change 2014*, 18 and 489.
3. Lynas, *Six Degrees*, 174.
4. Editorial, "How to Feed a Hungry World," *Nature* 466, no. 7306 (2010): 531–32.

There is no single class of "peasants," but rather several heterogeneous strata of waged laborers, sharecroppers, petty commodity producers, patriarchal family farms, and small-scale capitalist farmers. Some peasants work in quasifeudal modes of production, others on the peripheries of capitalism, while others hold land in common.[5] Agrarian social movements have emerged from these disparate strata to challenge aspects of capitalist agriculture, focusing particularly the retention and expansion of common rights and common ownership (against dispossession from the land) and against the use of transgenic crops. Among the more famous of these movements is the Landless Workers Movement (Movimento dos Trabalhadores Sem Terra or MST) in Brazil. A member of La Vía Campesina, it has organized land-takeovers and established growers' cooperatives while also engaging in wider social movement struggles.

Like the Zapatista National Liberation Army (EZLN) from Chiapas, Mexico—the MST not only recruit from the existing rural population, but also draw on a second generation of members returning to the land from the precarious, impoverished urban proletariat through what has been described as an attempted "re-peasantization" or "decolonial exodus" from the waged labor relation.[6] Like other LVC member-groups, they have consistently organized against multinational agribusiness, and particularly against genetically modified crops. They have occupied (and sometimes forced the closure of) GM research stations and farms and disrupted the distribution of genetically modified produce. For example, on International Women's Day in March 2008, 1,000 MST and LVC activists occupied a Monsanto facility and destroyed GM corn. In March 2016,

---

5. LVC defines a peasant as "a man or woman [sic] of the land, who has a direct and special relationship with the land and nature through the production of food and/or other agricultural products. Peasants work the land themselves, rely[ing] above all on family labor and other small-scale forms of organizing labor. Peasants are traditionally embedded in their local communities and they take care of local landscapes and of agro-ecological systems. The term peasant can apply to any person engaged in agriculture, cattle-raising, pastoralism, handicrafts related to agriculture or a related occupation in a rural area. This includes Indigenous people working on the land." [La Vía Campesina, "Declaration of Rights of Peasants—Women and Men" (2009), https://viacampesina.org/en/declaration-of-rights-of-peasants-women-and-men/.]

6. On re-peasantization, see Leandro Vergara-Camus, "The MST and the EZLN Struggle for Land: New Forms of Peasant Rebellions," *Journal of Agrarian Change* 9, no. 3 (2009): 365–91. For a post-autonomist reading, consider Paolo Virno's definition of exodus from the class relation as "a committed withdrawal, the recourse to force is no longer gauged in terms of the conquest of State power in the land of the pharaohs, but in relation to the safeguarding of the forms of life and communitarian relations experienced en route." [Paolo Virno, "Virtuosity and Revolution: The Political Theory of Exodus," in *Radical Thought in Italy: A Potential Politics*, eds. Paolo Virno and Michael Hardt (Minneapolis: University of Minnesota Press, 1996), 206.] However, one of his main examples is explicitly settler-colonial: North American workers fleeing wage labor for the frontier "in order to colonize low-cost land" (199). For Latin American land movements, especially in the case of the ELZN, this flight for autonomy from the state and the wage relation has an expressly decolonial aspect.

5,000 MST women destroyed GM pine and eucalyptus seedlings at a nursery owned by the Araupel corporation, one of the world's largest exporters of lumber and wood products.

These actions have incurred violent repression from state and private security agencies. In 2015, the Swiss-based agrochemical and biotechnology multinational company Syngenta was ordered by a Brazilian court to compensate the family of Valmir Mota de Oliveira for his murder by a private militia hired by Syngenta. Oliveira was assassinated during a protest in 2006 at a site where MST alleged the company was conducting illegal experiments on GM crops, in a zone with environmental protections, no less. Syngenta was also found responsible for the attempted murder of fellow MST activist Isabel do Nascimento de Souza.[7]

Grievances with GM crops fall into two basic categories: economic and technical. Prominent among the economic grounds for opposition is the effect on trade, with MST activists citing cross-contamination as leading to the loss of organic status and subsequent loss of premium prices for their produce. Another is the effect of seed monopolies on input prices. For example, consider Vandana Shiva's claim that the price of cotton seeds in India rose 71,000 percent after Monsanto cornered the market with their GM product. She argues seed patents represent an enclosure of the "genetic commons."[8] This aspect of monopoly and enclosure raises a third economic objection: demands for local control versus the globalized centralization of capital.

Technical grievances refer to the properties of specific transgenic crops, which are often designed to require high inputs of water, synthetic fertilizers, and insecticides. The monetary costs of these also constitute an economic grievance, though this high-input requirement is common to non-GM high-yield varieties of the Green Revolution too. Together, these factors can lock in the ecologically damaging aspects of industrial agriculture, such as workers' exposure to toxic chemicals, the death of pollinators, and the pollution of waterways by agrochemical runoff. Indeed, GMs like Monsanto's Roundup Ready lines were specifically designed to lock in monopolistic use of their accompanying agrochemicals.

7. This essay was written before the election of far right Jair Bolsonaro as President of Brazil in 2018. He has promised to designate the MST a "terrorist organization" and has vowed to purge Brazil of political enemies. Coupled with already ongoing violence against activists (e.g., Bahia state MST leader Márcio Matos was shot dead at his home in January 2018), this presents a terrifying picture for the MST and other progressive social movements. Despite the differences we have with some of the MST's approaches—detailed in this essay—we extend to them our full solidarity in their battle against capital's new fascist regime.

8. Vandana Shiva, "Why the Government is Right in Controlling the Price of Monsanto's Bt Cotton Seeds," *Scroll.In*, August 22, 2016, https://scroll.in/article/814476/why-the-government-is-right-in-controlling-the-price-of-monsantos-bt-cotton-seeds.

There is also the question of the "ecological arms race" triggered by pest-resistant GMOs. In 2014, for example, it was found that Bt maize—genetically engineered to produce insecticidal toxins derived from the Bacillus thuringiensis (Bt) bacterium—had created selective pressure for Bt-resistant pest species.[9] Such an arms race, in turn, further locks in monopoly control by big agribusiness, amplifying economic grievances.

The alternative to GMO dependent "food security" proposed by LVC and Shiva is "food sovereignty." This framework rejects the positioning of food as a commodity and instead values food producers over those who own the land on which food is produced or the patents through which food is grown. Food sovereignty movements have also called for the localization of "food systems" so that producers and consumers are closer together and for the localization of control over such systems. Finally, low external input production and harvesting methods are seen to be key to tackling climate change.[10]

A tension exists within the food sovereignty position, between the rejection of food as a commodity and the valorization of local markets which do not necessarily challenge the commodity form.[11] This tension reflects the heterogenous class composition of contemporary agrarian social movements. Insofar as a common interest has been found, it is that of petty commodity producers versus the big multinational capital of agribusiness. But, as these remarks have already intimated, opposition to the centralization of capital is not necessarily opposition to capitalist relations per se.[12] The problem with Syngenta, for example, is not that they are headquartered in Zürich but that they are capitalists.

9. Aaron J. Gassmann et al., "Field-Evolved Resistance by Western Corn Rootworm to Multiple Bacillus Thuringiensis Toxins in Transgenic Maize," *Proceedings of the National Academy of Sciences* 111, no. 14 (2014): 5141–46.

10. These are drawn from the "Six Pillars of Food Sovereignty," as laid down by the Nyéléni Declaration at the International Nyéléni Forum for Food Sovereignty in 2007 in Selingue, Mali, organized by peasant, environmental, and women's organizations. See La Vía Campesina, European Coordination (ECVC), "Food Sovereignty Now! A Guide to Food Sovereignty" (Brussels: European Coordination Vía Campesina, 2018), https://viacampesina.org/en/wp-content/uploads/sites/2/2018/02/Food-Sovereignty-A-guide-Low-Res-Vresion.pdf.

11. Shiva, *Soil Not Oil*, 117–19.

12. Centralization, as Marx points out, "is concentration of capitals already formed, destruction of their individual independence, expropriation of capitalist by capitalist, transformation of many small into few large capitals. . . . This is centralization proper, as distinct from accumulation and concentration." [Marx, *Capital* (1976), 777].

THE ORGANIC YIELD GAP

THE ORGANIC YIELD GAP

Shiva claims that "organic farming produces more food and higher incomes."[13] The latter part of this statement is very likely true, as many Western consumers are willing to pay a premium for unclear health and nutritional benefits.[14] Assessing whether organic farming could produce higher yields of food, however, is more complicated. A comprehensive meta-analysis published in *Nature* found "5 percent lower organic yields (rain-fed legumes and perennials on weak-acidic to weak-alkaline soils), 13 percent lower yields (when best organic practices are used), to 34 percent lower yields (when the conventional and organic systems are most comparable)."[15] Shiva argues precisely against such like-for-like comparisons, insisting that the alternative to high-input monoculture is low-input biodiversity (i.e., forms of polyculture). She presents a table which claims physical yield gaps of twenty-three percent, sixty-six percent, and seventy-five percent in favor of "biodiverse" versus "monoculture" production for three comparison sets.

The claim is not as implausible as it may sound. Polyculture practices can fill more ecological niches in the same space, and can therefore, in principle, boost physical yields while preventing weeds and limiting pests. On the other hand, polycultures can be less amenable to mechanical harvesting and so are more labor-intensive and thus less economically productive, from a reductive capitalist viewpoint. A 2015 study in the *Proceedings of the Royal Society* found that multi-cropping and crop rotation—two common agricultural diversification practices—substantially reduced the yield gap, but wherever these methods were also used in nonorganic production the yield gap expanded once again.[16]

We have not been able to find any meta-analysis or systematic review in the peer-reviewed literature that shows yield gaps in favor of organic agriculture, especially as a like-for-like comparison. However, individual studies—rather than systematic reviews or meta-analyses—do exist. One 2007 study found that "for most food categories, the average yield ratio was slightly <1.0 for studies in the

13. Shiva, *Soil Not Oil*, 105 and 107. In scientific terms "organic" farming is meaningless, since pesticides like DDT are organic compounds from a chemistry point of view, but colloquially the term has become well established as a vague synonym for "natural" (restrictions on pesticide use and GMO content).

14. A systematic review in 2012 found: "The published literature lacks strong evidence that organic foods are significantly more nutritious than conventional foods. Consumption of organic foods may reduce exposure to pesticide residues and antibiotic-resistant bacteria." Crystal Smith-Spangler et al., "Are Organic Foods Safer or Healthier than Conventional Alternatives?: A Systematic Review," *Annals of Internal Medicine* 157, no. 5 (2012): 348.

15. Verena Seufert, Navin Ramankutty, and Jonathan A. Foley, "Comparing the Yields of Organic and Conventional Agriculture," *Nature* 485, no. 7397 (2012): 229.

16. Lauren C. Ponisio et al., "Diversification Practices Reduce Organic to Conventional Yield Gap," *Proceedings of the Royal Society B: Biological Sciences* 282, no. 1799 (2015).

Contemporary Agriculture • 111

developed world and >1.0 for studies in the developing world."[17] As even the favorable studies don't find large superior organic yields, we are compelled to doubt Shiva's claims on this count. However, Shiva also makes an important argument against narrowly focusing on physical or economic yields, writing: "The promotion of so-called high-yielding varieties leads to the displacement of biodiversity. It also destroys the ecological functions of biodiversity. The loss of diverse outputs is never taken into account by the one-dimensional calculus of productivity. When the benefits of biodiversity are taken into account, biodiverse systems have higher output than monocultures."[18] This claim is plausible so long as output is understood broadly to include both negative and positive externalities. Notwithstanding the problems with attempting to price so-called "ecosystem services," estimates suggest they "contribute more than twice as much to human well-being as global GDP."[19] It seems plausible that the narrow economic efficiency of capital-intensive agriculture may disappear using a wider ecological calculus. For example, think of the poisoning of essential pollinators by insecticides and the depletion of soil fertility in synthetic fertilizer manufacture. Then, add in the manufacture, transport, and refrigeration of agrochemicals and equipment which results in large amounts of carbon emissions. Finally, take into account the possibilities for soil-based carbon sequestration and the pollution caused by agrochemical runoff. Such a broader view of agriculture would result in a different kind of comparison.

Additionally, most comparisons between conventional and organic agriculture are under optimal conditions—precisely the kind of stable, predictable growing conditions threatened by climate chaos. "Extrapolations of future crop yields must take into account the high likelihood that climate disruptions will increase the incidence of droughts and flooding in which case . . . OA [organic agriculture] systems are likely to out-yield CA [conventional agriculture] systems."[20] This is because conventional high-yielding varieties are optimized for fairly specific growing conditions, including high water inputs. For grains in particular, "temperatures over 30°C cause an escalating pattern of damage."[21] A warming (or more volatile)

17. Catherine Badgley et al., "Organic Agriculture and the Global Food Supply," *Renewable Agriculture and Food Systems* 22, no. 2 (2007): 86–108.

18. This point is acknowledged in the *Nature* study referenced above, in which it is noted that "yields are only part of a range of economic, social and environmental factors that should be considered when gauging the benefits of different farming systems." [Seufert, Ramankutty, and Foley, "Comparing the Yields of Organic and Conventional Agriculture," 231.]

19. Robert Costanza et al., "Changes in the Global Value of Ecosystem Services," *Global Environmental Change* 26 (2014): 152–58.

20. Donald W. Lotter, "Organic Agriculture," *Journal of Sustainable Agriculture* 21, no. 4 (2003): 72.

21. Lynas, *Six Degrees*, 157.

temperature and precipitation profile can severely interrupt the norms of such agricultural systems. As we write, severe and ongoing spring flooding in North America has meant that corn and soy farmers have been unable to plant their crops on a normal schedule. While it is possible that these crops will recover, futures markets are already skyrocketing, impacting the prices of corn- and soy-intensive foods like beef.

Alternative practices can have greater climate resilience. For example, inter-cropping a taller crop can provide cooling partial shade. Organically farmed soils are often more resistant to water and wind erosion. Ecological efficiency should also include the effects of shifting from feeding cattle (feedstock) and cars (bio-fuels) to feeding people. Measures to reduce massive food waste in the West must be taken.[22] It is precisely this reckoning with a multiplicity of incommensurable use-values which capitalist commodity production is incapable of fully addressing. It is this economic system, not conventional farming or genetic modification, that is the obstacle we must overcome.

CYBORG AGROECOLOGY

If we are not to concieve of "modern" technological formats of agroecology as in-herently damaging, how might we evaluate the choices presented to us by different forms of organizing social relations with and through the natural world? In a 1997 interview with Hari Kunzru, feminist theorist Donna Haraway offered the follow-ing thought experiment as a way of concieving the cascading interconnections un-captured by a simple 'nature versus technology' distinction:

> Imagine you're a rice plant. What do you want? You want to grow up and make babies before the insects who are your predators grow up and make babies to eat your tender shoots. So you divide your energy between growing as quickly as you can and producing toxins in your leaves to repel pests. Now let's say you're a researcher trying to wean the Californian farmer off pesticides. You're breeding rice plants that produce more alkaloid toxins in their leaves. If the pesticides are applied externally, they count as chemicals— and large amounts of them find their way into the bodies of illegal [sic] immigrants from Mexico who are hired to pick the crop. If they're inside the plant, they count as natural, but they may find their way into the bodies of the consumers who eat the rice.[23]

22. As a general index, only about ten percent of the energy consumed at one trophic level (cow eats grain) is available at the next trophic level (human eats cow). Therefore in general, shifting from meat to edible crop production increases the calories available to humans from a given area of land by an order of magnitude. [See Raymond L. Lindeman, "The Trophic-Dynamic Aspect of Ecology," *Ecology* 23, no. 4 (1942): 399–417.]

23. Hari Kunzru, "You Are Cyborg," *Wired Magazine* 5, no. 2 (1997): 1–7.

Haraway's point is not just to note that "natural" does not equal "good," although this is a fallacious appeal to nature all too common in environmental rhetoric. Rather, what is notable is her claim that *the distinction between "natural" and "artificial" does not withstand scrutiny.* As she puts it in "A Cyborg Manifesto," "the certainty of what counts as nature—a source of insight and promise of innocence—is undermined, probably fatally."[24] Accompanying the undoing of this binary, we suggest, must be the undoing of another series of absolute oppositions, including traditional vs. modern, fast vs. slow, living vs. nonliving, local vs. global, "organic" vs. "conventional." Such an understanding, coupled with an abolition of the commodity form, opens the door to a cyborg agroecology.[25]

The cyborg perspective is suspicious of organic holism, which seeks to locate a moral economy within a material one. This results in the series of binary oppositions described above which are operative in some contemporary advocacy of organic or sustainable farming. The idea that sustainable agriculture will or should lead to a revalorization of tradition, slowness or localism, for example, seems to elide the fact that it too is subject to power relations. Of course, organics under capitalism are not a worthwhile barometer—the problems with their bourgeois consumption politics and the cost and accessibility of organic labeling, for example, are numerous. Nonetheless, we need the ability to ask of such proposed changes basic political questions: Who wins? Who loses? Who decides? How do we know? Organic holism elides such politics.

We must overcome the corresponding binary between a traditional, natural, stable, life-giving, low-tech agriculture versus a modern, synthetic, dynamic, high-tech, capital-intensive arrgriculture. A story about two starkly opposed sides, it is of a piece with the self-image of colonial-capitalist modernity, in which good things like progress and modernization occur when dynamic Europe meets the people without history. Instead, we need an alternative theory of technology, one in which, as Jasper Bernes explains, "technology *is* nature, an organization of natural elements and powers."[26]

According to poltical scientist James C. Scott, well known for his extensive studies of food production in agrarian societies, "[t]he term 'traditional' . . .

---

24. Donna J. Haraway, *Manifestly Haraway* (Minneapolis: University of Minnesota Press, 2016), 12.

25. Many of Haraway's claims are by no means new or hers alone. Indeed, a rejection of many of these binaries is central to much Indigenous thought, while their dominance cannot be separated from Enlightenment thought's imbrication with colonialism.

26. Jasper Bernes, "The Belly of the Revolution: Agriculture, Energy, and the Future of Communism," in *Materialism and the Critique of Energy*, eds. Brent Bellamy and Jeff Diamanti (Edmonton and Chicago: MCM Prime Press, 2018), 335.

is a misnomer."[27] For Scott, so-called "traditional agriculture" is dynamic and plastic. It is the work of bricoleurs who make use of whatever materials and techniques are ready-to-hand, including selective use of scientific and technological tools. Practical skills and knowledge acquired through practice—which following Aristotle's classification of knowledges, Scott refers to as "mētis"—often surpasses formalized scientific knowledge since it is based on trial-and-error experimentation and tinkering. Bricoleurs may know that something works before they know how it works, albeit at higher risk of inferential errors (false positives/negatives). Rather than affirm either side of the traditional modern binary then, we should inquire into and seek to overcome the conditions under which it makes sense.

What conditions result in a world where labor-saving agricultural technologies are experienced as dispossession and urban poverty, rather than relief from drudgery and a multiplier of communal wealth? Under what conditions does formal scientific knowledge confront mētis producers on the land as the vanguard of capitalist dispossession? Could a more capacious set of knowledge practices adapt techniques from the formal sciences through bricolage? To what extent can traditional techniques be combined with modern technologies to boost yields, reduce toil, and maintain ecological relations at the same time?

One such practice is Integrated Pest Management (IPM), in which the use of chemical pesticides are allocated as "the position of the last resort in the chain of preferred options" for managing pests that threaten yield. Preferred options are more preventative approaches, including "organic farming, diversifying and altering crops and their rotations, inter-row planting, planting timing, tillage and irrigation, using less sensitive crop species in infested areas, using trap crops, applying biological control agents, and selective use of alternative reduced-risk insecticides."[28] Together, these practices seek to minimize rather than wholly eliminate the use of chemical pesticides.

27. James C. Scott, *Seeing Like a State: How Certain Schemes to Improve the Human Condition Have Failed* (New Haven: Yale University Press, 1998), 331.

28. J. P. van der Sluijs et al., "Conclusions of the Worldwide Integrated Assessment on the Risks of Neonicotinoids and Fipronil to Biodiversity and Ecosystem Functioning," *Environmental Science and Pollution Research* 22, no. 1 (January 1, 2015): 153.

IPM is one of many possibilities less readily apparent if we operate within the rhetorical binary between "life-giving" organic and "life-ending" nonorganic agriculture. Mētis practices of cyborg agroecology defy simplistic distinctions between modern and eco-friendly. For example, to minimize workers' exposure to harmful chemicals and to maximize the precision, speed, and safety of pesticide application, the selective use of drones for pesticide application might be worthwhile. Under capitalism, such experiments already underway will only serve counterrevolutionary ends, largely to displace workers and their power to strike. Consequently, we need to ask of agricultural technologies, as Bernes puts it, "how do revolutionary struggles beginning in the here and now find a way to meet their needs, survive, and grow, while producing communism?"[29] Such a reframing helps look at the path dependency and integrated nature of technological systems as well as push us beyond a simplistic binary of "before" and "after" the revolution.

In our contemporary context, it is for petty commodity producers that the strict delineation of organic agriculture makes the most sense. Their livelihoods require them to seek the highest market price for their commodities, and an organic label is more marketable than a "pesticides as a last resort" one. It is in this context that much of the activism of LVC, the MST, and Navdanya takes place. This is also a limit to the autonomy of groups like the EZLN. One can escape the wage relation locally, to an extent, but generalized commodity production and capitalist networks of logistics and distribution are only avoidable on a small scale for so long.

Shiva is right to emphasize the importance of self-organization for agrarian producers. But it is a mistake to translate this into championing local markets. Petty commodity production is repeatedly confronted by technology as the centralization of capital (i.e., the threat of being squeezed out by more highly capitalized rivals). Its perspective often seeks to promote commoning and cooperatives as alternative forms of production. Yet, these organizational forms can be enhanced through bricolage of newer technologies, including labor-saving ones. Agricultural bricolage could even include transgenics. If we were to treat GMOs as part of the genetic commons rather than enclosed, proprietary forms it would overcome many of the economic objections associated with GMOs.

In the disaster of a world warmed by 4°C or more—highly likely to arrive by 2100—we will be faced with an agricultural situation unlike any humans have ever encountered. One aspect of this will be challenges to traditional breeding

29. Bernes, "The Belly of the Revolution: Agriculture, Energy, and the Future of Communism," 333.

techniques. Climate-resilient, low-input strains conventionally crossbred with existing varieties will be needed to preserve biodiversity.

Could GMOs be useful to us amid disaster? GMOs are not as fast to develop as their boosters claim. A further problem with their production as commodities is that investment is driven by the prospect of monopoly rents through intellectual property to which conventionally unobtainable, uniquely useful phenotypes are a secondary concern. The point is largely moot under capitalist conditions, where the benefits of, for example, golden rice, are offset by the centralization of capital, the dispossession, impoverishment, and urbanization of the rural population, and the extension of the very market dynamics which ensure mass hunger amidst plenty.[30] Yet the potential advantages outside capitalist social relations are significant. Incorporating symbiotic nitrogen-fixing bacteria—currently limited to legumes—into staple crops through genetic modification, for example, would massively reduce dependence on synthetic fertilizers made with the energy-intensive Haber-Bosch process, which almost wholly relies on fossil fuels and has led to a planetary imbalance in the nitrogen cycle.

Cyborg agroecology should not be understood as having an inherent preference for high-tech solutions. From a cyborgian perspective, the assemblage peasant-ox-plough is no more or less a technonatural mesh than the assemblage AI-drone-GMO. The point is that bricolage practically appropriates whatever materials are to hand. For example, as the glaciers providing billions of beings with freshwater retreat, even maintaining traditional agriculture may well require desalination technology and knowledge of fluid mechanics to maintain irrigation. Yet a reprisal of "archaic" stormwater collection and distribution systems may also be able to play that role. Or, of course, the answer may lie with some combination of the two.

It is capitalist social relations which pit agricultural technologies against agricultural workers as well as scientific knowledge against mētis. It is this system of relations that makes local commodity production appear as the only alternative to global commodity production. They, not machines or transgenics per se, form the barrier to the kind of practices of bricolage necessary for avoiding the kind of

---

30. Production for the world market drives the financialization of agriculture, as farmers hedge price volatility with credit, futures, and options or are simply forced into debt, while the financial sector speculates on such assets. Such speculative dynamics were central to the global food-price crisis in 2007–2008. "Completely connected markets can generate feedback and loops which in turn create unexpected emergent behavior. . . . in increasing the autonomous flow of capital, directed by high frequency trading algorithms designed to expect static relationships, the markets create flash crashes, sudden shocks that shouldn't exist." [Alasdair, "Autonomisation, Financialisation, Neoliberalism," *libcom.org* (blog), January 7, 2013, http://libcom.org/blog/autonomisation-financialisation-neoliberalism-07012013.]

hunger inevitable under market dynamics. Agrarian social movements are surely essential to overcoming such a barrier, but the perspective of petty commodity production prevalent in those movements also forms part of that barrier.

If asked to point to the project of overcoming commodity relations, we would highlight communal approaches to production and distribution, the "food sovereignty" that LVC and others speak of. The promotion of gender equality, of nonhierarchical grassroots organization, and autonomous social reproduction through noncommodified food, housing, and healthcare provision cannot be dismissed as the parochial perspective of petty commodity producers. But agroecology's prospects of a wider overcoming will necessitate a communizing movement that encompasses urban struggles, refugee movements, and the selective repurposing of technologies bequeathed by capitalism, (re)inventing cyborg methods or reviving old ones, and unromantically finding what is adequate to the unfolding climate disaster.

JAMES O'CONNOR'S SECOND CONTRADICTION OF CAPITALISM

First published April 2014

Out of the Woods first started as a way of exploring the origins, legacies, and limits of ecological Marxism in a time of climate crisis. As our later essays demonstrate, our thought has shifted to better take into account social reproductive Marxist feminisms alongside anticolonial, antiborder, and antiracist thought. Nonetheless, our early introductory exploration and critique of James O'Connor's understanding of capitalist degredation of nature remains relevant to certain strands of eco-Marxism generally available in Monthly Review.

"The second contradiction of capital" is one of the most influential concepts in ecological Marxism. The outlines of this argument were first presented in 1988 by James O'Connor in "Capitalism, Nature, Socialism: A Theoretical Introduction," which appeared in the first issue of *Capital Nature Socialism*, a journal which he helped found. O'Connor's argument falls between Marxist orthodoxy, on the one hand, and a growing ecological awareness and the emergence of "new social movements" on the other. Responding to this context, it preserves certain core Marxist understandings while expanding them into new domains, like ecology, and for new subjects, such as feminist, environmental, and national liberation movements. In doing so, it relaxes other tenets of orthodox Marxism.

O'Connor draws inspiration from Karl Polanyi's argument in *The Great Transformation* that market relations come into conflict with social reproduction. He credits Polanyi's work as providing an alternative to dominant strains of socio-environmental thought: "bourgeois naturalism, neo-Malthusianism, Club of Rome technocratism, romantic deep ecologism, and United Nations one-worldism."[1] In what follows, we sketch out the "first" and "second" contradictions of capitalism. Then, we turn to engage with two of O'Connor's key ideas: an "underproductionist" theory of crisis and "the rebellion of nature" thesis.

A TALE OF TWO CONTRADICTIONS

The twenty-first century is a race between the first and second contradictions of capital. The first contradiction, central to orthodox Marxism, is between the forces of production and the relations of production. "Forces of production" include things like forms of labor, scientific knowledge, cooperation, and technology. The "relations of production" are the class relations arising out of private property in the means of production. In orthodox Marxism, this contradiction, via the labor movement, is held to a more or less linear progression towards centralized ("socialized") industry and more planning, by states or industrial-financial monopoly capitalists restructuring "into more transparently social, and hence potentially socialist, forms."[2] Socialism can then be understood as the workers (via a Communist Party) taking over the apparatus of centralized production and planning that capitalism bequeaths.

---

1. James O'Connor, "Capitalism, Nature, Socialism: A Theoretical Introduction," *Capitalism Nature Socialism* 1, no. 1 (January 1, 1988): 12–13.
2. O'Connor, "Capitalism, Nature, Socialism," 11.

O'Connor flirts with this orthodoxy but isn't satisfied with a unilinear, teleological history. Rather, he insists, "it has become obvious that much capitalist technology, forms of work, and the like, including the ideology of material progress, have become part of the problem not the solution."[3] He subsequently identifies a second contradiction which serves as a source of capitalist crisis. This is between the forces and relations of production combined and the *conditions* of production; or, broadly speaking, between political economy and the environment. "An ecological Marxist account of capitalism as a crisis-ridden system focuses on the way that the combined power of capitalist production relations and productive forces self-destruct by impairing or destroying rather than reproducing their own conditions," O'Connor writes.[4]

A CRISIS OF UNDERPRODUCTION?

Perhaps the most common crisis theories we encounter are underconsumptionist. The central claim of an underconsumptionist crisis theory is that it is profitable for individual capitalists to drive down wages, but when all capitalists do this consumer demand is suppressed, since poorer workers buy less. Thus capitalists are caught in a rationality trap. The solution is usually some form of social partnership, via the state, trade unions or both, to keep shortsighted capitalists from undermining aggregate demand and thus their own interests. For underconsumptionists, crisis is not inherent to capitalism. It can be avoided by more enlightened self-interest and social partnership, making capitalists realize that "decent wages" are a win-win.[5]

The more "properly" Marxist theory, by contrast, is one of overproduction.[6] This position argues that it is rarely in the interests of capitalists to pay higher wages, since it eats into their profits. It also points out an error in underconsumptionist thinking: consumer demand isn't the only demand because firms also consume; for example, by purchasing new machinery. Of course, firms only invest in the hope of future profits, so this creates a "grow-or-die" dynamic, but it means that underconsumption is a symptom rather than a cause of crisis.[7] Further, overproductionists note that firms invest

---

3. O'Connor, "Capitalism, Nature, Socialism," 16.

4. O'Connor, "Capitalism, Nature, Socialism," 22.

5. This is a simplification. Luxemburg's underconsumptionist crisis theory, for example, argues that crisis is inherent to capitalism, but can be deferred by expansion into noncapitalist regions via imperialism. But once the whole world was carved up by imperialist powers, the crisis would become inescapable.

6. Not all Marxists subscribe to this, however. For example, Michael Heinrich claimed to refute it (and argued that Marx himself came to reject the "law") in the controversial article "Crisis Theory, the Law of the Tendency of the Profit Rate to Fall, and Marx's Studies in the 1870s," *Monthly Review* 64, no. 11 (2013): 15.

7. For brevity's sake we are simplifying a large and contentious body of crisis theory here. For example, underconsumptionists counter that all demand is, ultimately, consumer demand.

in mechanized processes in order to gain an advantage over their rivals by becoming more productive. But as these productivity gains generalize across the industry, the rates of profit begin to fall. For overproductionists, this is the rationality trap: capitalists are compelled to become more productive to compete, but increased productivity and competition ultimately depress profit rates. Eventually, suitably profitable investments begin to dry up. Too much capital chases too few profits. Firms either engage in more speculative activity or withhold investment, the economy falters, some firms fail and others are devalued. The general devaluation of capital—crisis—helps restore the rate of profit, and then a new round of accumulation can begin.[8]

It is in this context that O'Connor proposes a different kind of crisis: underproduction.[9] Capital, he argues, increasingly dominates and incorporates its surroundings. This includes the natural environment, resulting in phenomena such as soil-nutrient depletion and deforestation. This requires the intervention of, for example, synthetic fertilizers and planned forestry.

It also has impacts on the social environment. Our health suffers from pollution and overwork, which necessitates expenses such as healthcare provision to maintain the workforce. In this way, nature itself becomes produced, so "that 'natural barriers' may be capitalistically produced barriers, that is, a 'second' capitalized nature."[10] Such a claim strikes us as profoundly correct. Indeed, as Marx put it in the *Grundrisse*, for capital, "every limit appears as a barrier to be overcome."[11]

Consequently, capital increasingly has to produce its own environment, and this represents an overhead cost at the systems level. A crisis of underproduction thus includes, but exceeds, a crisis of social reproduction, encompassing both social and environmental reproduction. O'Connor writes: "We can introduce the possibility of capital underproduction once we add up the rising costs of reproducing the conditions of production. Examples include: the healthcare costs necessitated by capitalist work and family relations; drug and drug rehabilitation costs;

---

8. Nothing in this account requires a terminal crisis, theories of which have more to do with political requirements: either to rationalize the "evolutionary socialism" of social democracy, or to compensate for a proletariat not living up to its ascribed revolutionary role.

9. O'Connor is developing an observation made by Marx: "The greater the development of capitalist production, and, consequently, the greater the means of suddenly and permanently increasing that portion of constant capital consisting of machinery, etc., and the more rapid the accumulation (particularly in times of prosperity), so much greater the relative over-production of machinery and other fixed capital, so much more frequent the relative under-production of vegetable and animal raw materials." [Karl Marx, *Capital: A Critique of Political Economy*, Volume III, trans. David Fernbach (London: Penguin Classics, 1981), 770.

10. This argument parallels the move from formal to real subsumption of the labor process, which is Marx's account of how capital remakes production in its image—by deskilling, organizing production lines and global value chains etc. [See Karl Marx, "Economic Manuscripts: Marx's Economic Manuscripts of 1861–63," accessed April 26, 2019, https://marxists.catbull.com/archive/marx/works/1861/economic/ch37.htm.]

11. Karl Marx, *Grundrisse* (London: Penguin, 1973), 408.

the vast sums expended as a result of the deterioration of the social environment (e.g., police and divorce costs); the enormous revenues expended to prevent further environmental destruction and to cleanup or repair the legacy of ecological destruction from the past."[12]

In light of the massive costs associated with mitigating and adapting to climate change, this line of argument seems highly salient. It also provides a framework to understand the contemporary discourse around "ecosystem services": the framing that capitalists put on the benefits provided by ecosystems. Once natural cycles are perturbed and/or replaced by political-economic ones, the price of nature's "free gifts" is revealed in rising financial costs. Examples could include the use of (energy intensive) synthetic fertilizers to compensate for declining soil fertility or the need to hand-pollinate crops where intensive pesticide use and loss of hedgerow habitats have wiped out pollinators. For O'Connor, this dynamic feeds back into economic crisis. He notes that "no one has estimated the total revenues required to compensate for impaired or lost production conditions and/or to restore these conditions and develop substitutes . . . all unproductive expenses from the standpoint of self-expanding capital."[13]

THE REBELLION OF NATURE

O'Connor explains the rise of new social movements as a response to the crises of underproduction. He suggests that "the combination of crisis-stricken capitals externalizing more costs and the reckless use of technology and nature for value realization in the sphere of circulation must sooner or later lead to a 'rebellion of nature,' that is, to powerful social movements demanding an end to ecological exploitation."[14] This stretches the concept too far. First, it is politically problematic to frame feminist, Indigenous, and antiracist movements as a unified 'rebellion of nature.' Second, while capital's devastation of ecologies has solicited a wide variety of struggles and movements, there is nothing inevitable about the rise of a powerful, unified class movement. Ecocide provides the grounds for grievance, but grievance is only one of the conditions for a movement.

---

12. O'Connor, "Capitalism, Nature, Socialism," 23. Here, O'Connor follows Marx's insight that "[t]he maintenance and reproduction of the working class remains a necessary condition for the reproduction of capital. But the capitalist may safely leave this to the worker's drives for self-preservation and propagation." [Marx, *Capital* (1976), 718.]

13. O'Connor, "Capitalism, Nature, Socialism," 23. Since O'Connor wrote this there have been numerous attempts to forecast the total economic cost of climate change.

14. O'Connor, "Capitalism, Nature, Socialism," 28.

In the case of climate change, we encounter a further problem. With a polluting factory or mine, the effects are often felt most harshly in the immediate vicinity, or at least in specific localities, e.g., downriver, and the origin is relatively easily identified. But climate change is localized in neither space nor time. Aside from the impossibility of attributing the effects of climate change to any specific greenhouse gas emissions, the inertia of the climate system means the emissions that caused this warming happened decades ago. Warming from recent emissions is yet to be felt, even if we were to stop all emissions today. How can a movement coalesce against a threat that is so diffusely and enormously distributed across space and time? We are not saying this is impossible, only that it cannot be taken for granted that "sooner or later" powerful social movements will cohere.

In a sense, this is no different to more narrowly defined class struggle. Capitalism generates grievances daily. A small percentage of these flare up into local struggles. A percentage of these then broadens out or links up into wider movements. A percentage of such movements escalates to shake the whole social order. Fragmentation and multiplicity are the norm. Ecological struggles seem to follow the same pattern. Indeed, O'Connor insists that "issues pertaining to production conditions are class issues (even though they are also more than class issues)."[15] Climate change will certainly escalate the grievances, with crop failures, rising food prices, and displaced populations to name but three. As with crises related to the first contradiction, there can of course be no guarantee that political shifts will be positive. Indeed, when he writes that "atmospheric warming, acid rain, and pollution of the seas will make highly social forms of reconstruction of material and social life absolutely indispensable," we can only agree. It is not, however, inevitable. The intensification of capitalist social relations through fascism is also highly possible.

15. O'Connor, "Capitalism, Nature, Socialism," 32.

MURRAY BOOKCHIN'S LIBERATORY TECHNICS

First published March 2014

Since our introductory exploration of Bookchin's social ecology approach, his popularity has exploded worldwide as a thinker whose influential ideas have most recently been popularized by the Kurdish resistance, and contemporary "radical municipalism" in Europe and North America.[1] Though we have misgivings with taking Bookchin's thought as a roadmap (difficult in itself as he changed paths several times), here we explore an underappreciated aspect of social ecology: its pragmatic stance towards technics. Though the benefits of the deliberative approach here are sometimes overstated, we suggest such tools might be part of the process of any ecological transition.

---

1. Joris Leverink, "Murray Bookchin and the Kurdish Resistance," *ROAR Magazine*, August 9, 2015, https://roarmag.org/essays/bookchin-kurdish-struggle-ocalan-rojava/; Debbie Bookchin, "Radical Municipalism: The Future We Deserve," *ROAR Magazine*, July 21, 2017, https://roarmag.org/magazine/debbie-bookchin-municipalism-rebel-cities/.

Murray Bookchin (1921–2006) was a pioneer of radical ecological thought and a working-class autodidact. His political trajectory took him from Stalinism (joining a youth movement at age nine), through Trotskyism and anarchism, to a unique development of libertarian ecological thought he variously termed "social ecology," "libertarian municipalism," and "communalism." This brought him to a late engagement with reformist local politics in Vermont.

Unlike so many thinkers on the revolutionary Left in the sixties and seventies, Bookchin was able to clearly identify the shortcomings of both capitalist development and state socialism. He understood both systems were not only political and economic failures but also creative of ecological and technological crises. Seeking a way out of this bind he attempted not to resolve the contradiction between state socialism and bureaucratic capitalism (as with the administrative and technocratic Green Parties of the eighties). Instead, Bookchin located ecological liberation as a project directly tied to the release of human capacities through certain technologies.

Bookchin draws on Aristotle's notion of *techné* [technique], as encompassing a web of social relations and ethical principles, including finitude and limits. Techné includes a wide range of human activity, such as art and ethics, and "thus encompasses not merely raw materials, tools, machines, and products but also the producer—in short, a highly sophisticated subject from which all else originates." Capital, as "modern industrial production," treats technology as an entity separate from its social context, and understands its sole purpose as limitless production.[1] Bookchin describes this transformation by stating that "once societal constraints based on ethics and communal institutions were demolished ideologically and physically, technics could be released to follow no dictates other than private self-interest, profit, accumulation, and the needs of a predatory market economy."[2] A liberatory technics would thus require the re-embedding of technology in webs of communal social relations and ethics, so that it could once again uphold other kinds of social relations and ethics.

Bookchin's philosophy of technology can be best understood as a response to the limits of three other modes of societal organization: organic, class, and Promethean.

1. Murray Bookchin, *The Ecology of Freedom: The Emergence and Dissolution of Hierarchy* (Palo Alto, CA: Cheshire Books, 1982), 222. This view is reminiscent of, and parallel with, French and Italian autonomist Marxists of the seventies, who were especially influenced by the publication and translation of the so-called "Fragment on Machines" from Marx's *Grundrisse*. Bookchin argued that the Aristotelian image of technics anticipated that of Marx.
2. Bookchin, *The Ecology of Freedom*, 254.

By the first of these, Bookchin does not mean the conception of the social as an organism, or organic whole, but rather those pre- and noncapitalist forms of social organization which are internally composed in solidarity or kinship with the ecological world. In such societies, he finds a kind of egalitarian and nonhierarchical mode of being, which foster "the conceptual means for functionally distinguishing the differences between society and nature without polarizing them."[3] Though this is inspirational to Bookchin, he has no desire to return to a romanticized hunter-gatherer or agrarian subsistence mode of living, as in deep ecology, primitivism or anticivilization.[4] Indeed, he rubbishes the idea that humans can avoid interfering with nature since, as biological entities with nutritional requirements, we are part of it.

For Bookchin, organic societies are marked by a potentiality for hierarchy that they find difficult to ward off. This is not due only to the emergence of classes, but prior historic consolidations of power and divisions of labor between men and women, peasants and lords, old and young, priests and subjects, and eventually, the state and society. It is only through the consolidation and hardening of such barriers that a full-fledged class society emerges.[5] Historically, this consolidation was largely driven by European colonialism and was not teleological, but "occurred unevenly and erratically, shifting back and forth over long periods of time."[6]

The profit motive operating in class society limits human creativity to that which can be commodified. This favors large-scale industry, hierarchical management, and the subsumption of the labor process into the deadening, alienated toil of the assembly line. An autoworker in his youth, Bookchin experienced repetitive labor as particularly loathsome for its subjugation of social potentiality. Such toil allows no creativity, and so it cannot produce total well-being. Echoing Kropotkin, Bookchin saw that despite the visible abundance of mass production, scarcity had to be artificially maintained:

---

3. Bookchin, *The Ecology of Freedom*, 254.

4. With the insight of critical Indigenous thinkers, we might complicate Bookchin's anthropological assessment (and ultimate dismissal) of "organic societies." For Bookchin, one central danger of organic societies are that they lack a distinction between domination and liberation. Freedom simply isn't a meaningful category, he supposes (see *The Ecology of Freedom*, 44, 140). Yet contemporary revolutionary Indigenous movements are not only fully capable of enacting liberation, but do so fully within their own ontologies. See, for example: Coulthard, *Red Skin, White Masks*; Estes, *Our History Is the Future*; Leanne Betasamosake Simpson, *As We Have Always Done: Indigenous Freedom through Radical Resistance* (Minneapolis: University of Minnesota Press, 2017).

5. Bookchin, *The Ecology of Freedom*, 87.

6. Bookchin, *The Ecology of Freedom*, 6.

Let us consider a factor that has played an important ideological role in shaping contemporary society: the 'stinginess' of nature. Is it a given that nature is stingy and that labor is humanity's principal means of redemption from animality? In what ways are scarcity, abundance, and post-scarcity distinguishable from each other? Following the thrust of Victorian ideology, do class societies emerge because enough technics, labor, and 'manpower' [sic] exist so that society can plunder nature effectively and render exploitation possible, or even inevitable? Or do economic strata usurp the fruits of technics and labor, later to consolidate themselves into clearly definable ruling classes?[7]

Bookchin would firmly and consistently answer affirmatively in the direction of the latter. As he pithily put it: "A century ago, scarcity had to be endured; today it has to be enforced."[8] It is clear that this is largely an accomplishment of the state, which is able to solve crises of overproduction through spatial fixes. Consider, for example, the IMF's opening of trade borders to enable the dumping of excess dried milk on the Global South in the mid-eighties.

Alternatively, capitalists may tend toward the production of disposable goods: "the shabby, huckster-oriented criteria that result in built-in obsolescence and an insensate consumer society."[9] Bookchin's third position—a "Promethean, often crassly bourgeois" Marxism—took precisely such industrial development to be a necessary aspect of a transition to communism.[10] This version of Marx's philosophy of technology is familiar to the point of being stereotypical today and has been revised and debated endlessly by ecosocialists and materialists alike. At Bookchin's time of writing, industrial Marxism was visceral and depressing. Soviet and Maoist development models were taken to be the real enactment of Marx's philosophy, with horrendous results for both people and their environments. Bookchin thus returns to Marx to critique the idea that Nature, as a realm of want and necessity, is merely something to be conquered, appropriated, and subjugated by boundless productive forces. The problem of scarcity and ecological degradation alike are taken to be merely a matter of unequal distribution. The result of such an assessment of Man and Nature is a "sharp disjunction . . . between society, humanity, and

7. Bookchin, The Ecology of Freedom, 66.

8. Murray Bookchin, Post-Scarcity Anarchism (Oakland: AK Press, 2004), 5.

9. Bookchin, Post-Scarcity Anarchism, 77.

10. Bookchin, The Ecology of Freedom, 226. At some point between his work in the 1960s (collected in Post-Scarcity Anarchism) and his magnum opus The Ecology of Freedom (first published in 1982), Bookchin seems to have read Adorno and Horkenheimer's Dialectic of Enlightenment. This appears to have dampened his earlier enthusiasm for scientific knowledge and productive technologies. In the sixties he enthusiastically advocated full automation and focused on the technologies which made this possible. By the eighties, his emphasis had shifted to the social matrix within which technology operates. We see this evolution as primarily a shift in emphasis, not a distinct break.

'needs' on the one side, and nature, the nonhuman living world, and ecological ends on the other."[11]

Out of these different and overlapping modes of relating with nature and technics, Bookchin developed another, utopian possibility, which he called "ecological society." Here, technics would not suggest an instrumental approach to nature but rather work in cooperation with and nourish ecological processes. Instead of the technological options posed by organic societies, class society and Promethean Marxism, Bookchin advocates the use of what he calls "liberatory technology," which involves automating repetitive work and moving towards modularity and self-assembly so that large-scale automated production facilitates creative, flexible, and local assembly. On a whole, however, the objective is is a planetary-wide liberation from degrading work.

Three core heuristics guide Bookchin's notion of liberatory technology: (1) liberation from toil; (2) the creation of collective social relations capable of symbiosis with ecologies; and (3) amenability to face-to-face, assembly-based direct democracy.

Liberation from toil is not merely quantitative: it is more than a matter of reducing hours of hard, repetitive labor while increasing free time. In fact, for Bookchin, this very distinction implies the persistence of alienated labor. Rather, the reduction of toil also has a qualitative dimension. Bookchin turns here to the traditions of peasant life, noting that "[t]heir most striking feature is the extent to which any kind of communal toil, however onerous, can be transformed by the workers themselves into festive occasions that serve to reinforce community ties."[12]

Bookchin's understanding of technology as imbricated with social relations leads him to technologies that can support the reproduction of libertarian, communal relations, operating in line with wider ecological processes. He argues that "the real issue we face today is not whether . . . new technology can provide us with the means of life in a toil-less society, but whether it can help to humanize society, whether it can contribute to the creation of entirely new relationships."[13] This remaking of social relations seeks "a balanced relationship with nature" through communities which live in a "symbiotic relationship with their environment."[14]

In direct contrast to the ways capitalist technologies separates us from natural processes to keep nature at bay and to dominate it, liberatory technology can

---

11. Bookchin, *The Ecology of Freedom*, 232.

12. Bookchin, *The Ecology of Freedom*, 255. Lest this be mistaken for nostalgia, Bookchin is clear that "a restoration of . . . peasant agriculture . . . is neither possible nor desirable." [Bookchin, *Post-Scarcity Anarchism*, 137.]

13. Bookchin, *Post-Scarcity Anarchism*, 48–9.

14. Bookchin, *Post-Scarcity Anarchism*, 58 and 83.

inculcate a sense of mutual dependence on nature into the fabric of everyday life. This is not simply a manner of inverting alienation, because the resulting relationship would "add a sense of haunting symbiosis to the common productive activity of human and natural beings."[15] Unlike much of the green movement, Bookchin sees no inherent beauty in small-scale, "soft" or "appropriate" technologies alone. Rather, a liberatory technology would not be judged based on size or type, but how it augments and supports liberatory social formations.

Finally, Bookchin stresses the need for technologies to be compatible with face-to-face, assembly-based direct democracy. He elaborates a critique of the council system as expounded by many Marxists, where decision-making and administrative functions are fused (as in Marx's account of the Paris Commune). Rather, to avoid the overbearing nature of representation and bureaucracy, he stresses that decisions must be made in assemblies, while council bodies should be strictly limited to administering the assembly's decisions, and made up of recallable delegates. Importantly, he stresses that a collective in an ecological society "is not merely a structural constellation of human beings, but rather the practice of *communizing*."[16] That is, the ongoing production of freedom-through-commons is a set of relationships continually reproduced through social and ecological interactions.

Bookchin is often caricatured as a naive localist. Indeed, his claims regarding the importance of federating autonomous communities sometimes seems an afterthought, especially given the spatial and scalar challenges of such decision-making. This is especially concerning given the challenges posed by climate change. "Local autonomy" clearly cannot extend to local people burning local coal for local energy. Our warming world makes clear that we live in a global commons, and makes federation one of the most difficult political problems of our time.

Localized autonomy must be understood accordingly. Even E.F. Schumacher, the green economist and author of *Small is Beautiful* (often read as romanticizing the local), calls for the principle of "subsidiarity" to be adopted, whereby decisions are made at the most local level possible.[17] Unlike thinkers such as Schumacher and "glocalist" Greens, however, Bookchin accepts that wider, revolutionary changes needed are needed. For example, the elimination of "the money economy, the state power, the credit system, the paperwork and the policework required to hold society

---

15. Bookchin, *The Ecology of Freedom*, 263. [Emphasis added.]
16. Bookchin, *The Ecology of Freedom*, 263. [Emphasis in original.]
17. E. F. Schumacher, *Small is Beautiful: Economics as if People Mattered* (New York: Harper, 1973). Anarchist political theorist Uri Gordon refigures subsidiarity as "the principle that people should have power over an issue in proportion to their stake in it." [Uri Gordon, "Anarchism and Nationalism," in *Brill's Companion to Anarchism and Philosophy* (Leiden: BRILL, 2018), 209.]

in an enforced state of want, insecurity and domination."[18] Bookchin seems to anticipate more recent critiques of the individualism implied by autonomy (*auto* + *nomos* means "self-law") and "autonomy + anti-State violence = revolutionary movement."[19] It is this tendency to totalize (even as localities are empowered), and this insistence on destroying the present state of things, that leads us to prefer the label "communist" to Bookchin's "communalism" or "democratic confederalism."

Bookchin does not have a huge amount to say about the process by which liberatory technologies supplant capitalist ones, other than by direct action and revolution. However, his insistence that "direct democracy is ultimately the most advanced form of direct action" opens the door to reformism.[20] We wonder if he prioritizes process over antagonism, and hence potentially participation in the local state electoral apparatus ahead of collectively organized class struggle. This position seems to become more explicit in his later writings. Perhaps Bookchin's battle with his own orthodox Marxist past sometimes led him endorse a collective subject determined by locality rather than alliances and intersections based on material identities (i.e., identity positions determined by and resistant to power).

The words he closes *The Ecology of Freedom* with remain remarkably prescient three decades later:

> Our technics can be either catalysts for our integration with the natural world or the chasms separating us from it. They are never ethically neutral. . . . The rewards we can glean from the wreckage they have produced will require so much careful sifting that an understandable case can be made for simply turning our backs on the entire heap. But we are already too deeply mired in its wastes to extricate ourselves readily. . . . In the end, however, we must escape from the debris with whatever booty we can rescue. . . . The means for tearing down the old are available, both as hope and as peril. So, too, are the means for rebuilding. The ruins themselves are mines for recycling the wastes of an immensely perishable world into the structural materials of one that is free as well as new.[21]

Bookchin's philosophy of technology, despite our reservations with some aspects of it, remains an important point of departure for us. It informs critiques of capitalist technological fixes to climate change that leave the underlying exploitative social relations untouched or deepened while also providing an heuristic framework for thinking about technology's emancipatory potential.

18. Bookchin, *Post-Scarcity Anarchism*, 82.
19. Gilles Dauvé, "A Contribution to the Critique of Political Autonomy," *libcom.org* (blog), October 26, 2008, http://libcom.org/library/a-contribution-critique-political-autonomy-gilles-dauve-2008.
20. Bookchin, *The Ecology of Freedom*, 339.
21. Bookchin, *The Ecology of Freedom*, 345–47.

ORGANIZING NATURE IN THE MIDST OF CRISIS

JASON W. MOORE'S
CAPITALISM IN THE WEB OF LIFE

First published January 2016

If we are not to understand 'nature' as an externalized organic whole despoiled by man, how should we understand its supposed common sense in ecological politics around the world? Jason W. Moore suggests this is in part because the split between 'nature' and 'artifice' (technics, humanity, culture) is a functional ideology for contemporary capitalism. Exploring the implications of this political economy, we find that the role of political struggle is crucial to emphasize. When capitalist ideology is presented as all pervasive, we run the risk of occluding the multiple pathways out of our contemporary crisis—especially those practiced by Black and Indigenous peoples, migrants, and people of color worldwide.

A common and widely accepted position on the Left is that capitalism destroys nature. It is undeniable that the imperatives of capital accumulation have resulted in destruction to myriad and complex sets of ecological relationships, flows, and life-forms. Capital relies upon ecological systems and flows for support or appropriation, while at the same time, it treats other parts of nonhuman nature as sinks or garbage dumps. This system is untenable, as James O'Connor has argued. Ecosocialist thinkers, such as John Bellamy Foster, argue that ecology is *against* capitalism and these constitute two separate worlds in dialectical conflict with each other. But can we really understand the natural world as something pure and pre-existing our current catastrophe? Or is the understanding that nature is something separate in fact crucial to the functional ideology of capital?

Historian Jason W. Moore's *Capitalism in the Web of Life: Ecology and the Accumulation of Capital* problematizes the widely shared view that "nature," "the environment" or "ecological systems" could constitute a separate and unique part of existence outside of capitalism and are incorporated into capital or devalued by it. Moore's "world-ecology" approach proposes that nature is always *in* capital. Capital does not appropriate natural resources from some outside realm but is organized internally to "see" the world as a reserve of resources and flows.

Likewise, capitalism is always *in* historical natures. It emerges within ecological contexts. Such dual implication (capital-in-nature/nature-in-capital) does not mean that there is no outside. Instead, it suggests that when considered historically or philosophically, nature, ecology, and environment cannot be understood as exteriorities. Capital "does not act upon nature but develops through the web of life."[1] In Moore's view, the modern world-system is a capitalist world-ecology, joining the accumulation of capital, the pursuit of power, and the production of nature in dialectical unity. Nature conditions capitalist accumulation and is produced historically by capitalist relations. In Moore's clearest formulation, "capitalism is not an economic system, it is not a social system, it is *a way of organizing nature*."[2] This allows us to see how dependent both accumulation and the exploitation of labor are on the appropriation and reproduction of "cheap nature," Moore's designation for energy, raw materials, food, and labor-power.[3] These are *rendered* "cheap" in the

---

1. Jonah Wedekind, "Jason W. Moore: Political Ecology or World-Ecology?," January 12, 2016, https://undisciplinedenvironments.org/2016/01/12/jw-moore-politicalecology-or-worldecology/.

2. Jason W. Moore, *Capitalism in the Web of Life: Ecology and the Accumulation of Capital* (London and New York: Verso Books, 2015), 2. [Emphasis in original.]

3. Moore (in collaboration with Raj Patel) has recently added cheap money, cheap care, and cheap lives. [Raj Patel and Jason W. Moore, *A History of the World in Seven Cheap Things: A Guide to Capitalism, Nature, and the Future of the Planet* (Oakland: University of California Press, 2017).]

sense of "the periodic, and radical, reduction in the socially necessary labor-time of these Big Four inputs."[4]

Moore's *Capitalism in the Web of Life* is a monumental attempt to follow the consequences of this view, and deserves praise for its many meticulous arguments. While we appreciate Moore's synthetic approach, he fails to explain why the nature/society split continues and how it might be effectively dismantled. Answering these questions, we believe, is the key to unlocking an epochal crisis in capitalism. The crisis won't come from nature alone. In short, capitalism won't end without us. We have to make it so.

Dominant, common-sense understandings position "nature" and "society" as separate entities that interact in various ways. This is a legacy of Western European Enlightenment thinking, which has been spread and imposed by colonialism.[5] Moore calls this nature-plus-society thinking "Green Arithmetic." Everyday life under capitalism generally primes us to think of nature as our resource and, conversely, our impacts as ecological footprints. The Left's version of this is evident in the subtitle of Naomi Klein's latest book *This Changes Everything: Capitalism vs the Climate*, while a neoliberal version is apparent in The Ecomodernist Manifesto's proposal to "decouple" capitalism from the environment through intense technological development, thus "making more room for nature."[6] Broadly speaking, people too often act as if nature is one of three things: a tap of resources, a sink for waste, or an external limit for economic growth. Rarely do we think of ourselves and our social and economic organization as aspects of nature.

For Moore (as for many others), this division cannot hold. Yet shaking it off will be hard. The dualistic opposition of nature and society makes it difficult to think of "Wall Street as a way of organizing nature," as one of Moore's provocative formulations puts it. In place of Green Arithmetic, Moore proposes the notion of the *oikeios*. Borrowed from the ancient Greek botanist Theophrastus, this is a contraction of *oikeios topos* [favorable place] and is used to name the relationship between a plant and its environment. Here, "the *oikeios* is a multi-layered dialectic, comprising flora and fauna, but also our planet's manifold geological and biospheric configurations, cycles, and movements. . . . From the perspective of the oikeios,

---

4. Moore, *Capitalism in the Web of Life*, 53.

5. The cosmologies of many Indigenous peoples across the planet are not necessarily premised on this split, which is one among many reasons why colonialism has deliberately sought to eradicate Indigenous knowledge and replace it with European ontologies. For the full implications of this difference, see especially Watts, "Indigenous Place-Thought and Agency Amongst Humans and Non Humans (First Woman and Sky Woman Go On a European World Tour!)."

6. John Asafu-Adjaye et al., "An Ecomodernist Manifesto" (2015), http://www.ecomodernism.org/manifesto-english.

civilizations do not 'interact' with nature as resource (or garbage can); they develop through nature-as-matrix."[7] This understanding leads Moore to argue that "nature can neither be saved nor destroyed, only transformed."[8] Such a thermodynamic formulation strikes us as profoundly contrary to conventional environmentalist wisdom. It allows Moore to analyze oil next to labor, the household next to super-weeds, contemporary climate change next to sixteenth-century sugar plantations. Each of these is equally historically and relationally produced by and through capitalism, and in turn conditions capitalism's historic and future ability to extract value from its natures (including human natures).

Perhaps Moore's most important application of *oikeios* is to tease out the relationship between paid labor power (exploitation) and the appropriation of unpaid work/energy—what might be framed as "women [and nonbinary people], nature, and colonies," to adapt Maria Mies' formulation.[9] If we accept that nature is not a timeless background to capitalism, but instead that "historical natures" are produced by and products of modes of production, then it becomes increasingly clear that historical natures and their reproduction are not incidental to accumulation. Capital needs new sites of enclosure. Crops are turned into commodities. Genetic sequences get patented and become the private property of agriculture corporations. Even carbon dioxide, a base element, is converted into carbon-dioxide emissions permits to be traded on markets. Natures are the condition of capital's possibility via a dialectic of capitalization and appropriation.

Capitalization—of labor power, of individual products, of land and territory—is the moment of turning nature into capital, bringing it within the sphere of directly market-mediated social relations. It tears out individual units from the web of life, creating discrete entities to be accounted for rather than relational meshes. Gold, iron, and silver have to be separated from slag. Whale oil has to be rendered from blubber attached to living creatures in a vast ocean. Labor productivity has to be separated from leisure. But capitalization, in which commodities exist as part of the recognized economy, is fundamentally reliant on appropriation. Appropriation refers to the reliance of that economy on unpaid labor and nature. The mine relies on appropriating its surroundings with toxic waste—"appropriation through pollution," in Michel Serres' words.[10] Whale oil relies on the ecological relations

7. Moore, *Capitalism in the Web of Life*, 4.

8. Moore, *Capitalism in the Web of Life*, 45.

9. Maria Mies, *Patriarchy and Accumulation on a World Scale: Women in the International Division of Labour*, 2nd Edition (London: Zed Books, 1998), 77.

10. Michel Serres, *Malfeasance: Appropriation Through Pollution?*, trans. Anne-Marie Feenberg-Dibon (Stanford: Stanford University Press, 2011).

of the ocean (not to mention the rest of the whale's body) to support and produce blubber. And capitalized labor-power relies on unpaid carework and housework to reproduce its worker.

With its important albeit too narrow focus on exploitation through (under-) paid labor, orthodox Marxist theory has been unable to account for the extra-economic appropriation of unpaid labor and nature. Moore, by contrast, centers the role of appropriation in the production of value, writing that "value as abstract labor [that is, labor oriented to the production of a commodity] cannot be produced except through unpaid work/energy." It follows that "these movements of appropriation must, if capital is to forestall the rising costs of production, be secured through extra-economic procedures and processes."[11]

Through this formulation, Moore seeks to show that the capitalist economy has always relied upon appropriation—of Indigenous lands, of slave bodies, of women's reproductive labor—to supplement and reproduce the conditions for economic exploitation. Slavery is central to accumulation, not a pre-capitalist formulation. European settler colonization and "frontier" adventurism is not a thing of the past but is necessary for ongoing processes of accumulation, both in "traditional" forms and through the opening up of new frontiers across scales from the microbial to the interplanetary (and from genetics to space exploration).

To further capture the importance of appropriation to capital, Moore introduces the concept of ecological surplus: "the ratio of the system-wide mass of capital to the system-wide appropriation of unpaid work/energy."[12] This involves the capacity of capital to appropriate sufficient unpaid work to allow for the possibility of accumulation at an acceptable rate of profit. From this formulation, Moore can pose a whole set of further historical questions. For example: does the ecological surplus have a tendency to fall (become exhausted, depleted, toxic, etc.)? If so, is capitalism reaching a crisis point? Is this crisis *developmental* (leading to a new restructuring of capital accumulation) or *epochal* (threatening accumulation as such)?

Moore argues that without a sufficient share of ecological surplus relative to capitalization, accumulation is not possible. It is as if capital, when forced to pay the full costs of production, could advance only at the most meager rates of profit. Yet over the course of each long wave cycle of accumulation there is a natural tendency for this surplus to fall. He suggests four reasons for this. Firstly, there is "wear and tear on the *oikeios*."[13] Secondly, the mass of capitalized nature tends to rise faster

11. Moore, *Capitalism in the Web of Life*, 65 and 67.
12. Moore, *Capitalism in the Web of Life*, 95.
13. Moore, *Capitalism in the Web of Life*, 97.

than new unpaid work can be appropriated. Thirdly, there is a contradiction between the reproduction times of nature and capital—capital must always strive to accelerate, while nature is limited in how fast it can reproduce. For example, fossil fuels can be extracted and burned far faster than geological processes can create them or remove their carbon from the atmosphere. Finally, as the wastefulness of capital increases, waste accumulates and grows more global and more toxic over time. Already-existing nuclear waste will need to be closely monitored for longer than human beings have existed so far, presenting a timescale difficult to politically organizing around. Toxic e-waste dumps such as Agbogbloshie in Ghana are home to tens of thousands of people, receiving concentrated waste streams from all over the world. And of course, greenhouse gases keep accumulating in the atmosphere, pushing the planet's climate towards chaos. These examples demonstrate how one might understand a world-ecology in which the appropriation of ecological surplus becomes more difficult and costly for capital to uphold.

The externalities rise, increasingly imposing what Moore calls "negative value."[14] For example: genetically modified crops are preyed upon by superweeds, which can only be countered by increased labor input or increased amounts of toxic herbicides. Agricultural productivity declines because of the effects of climate change. Pollution from unconventional sources of energy like tar sands and hydrofracking is now promptly identified and firms come under pressure to be accountable for these externalities. Consequently, costs inevitably increase and form an impediment to accumulation.

Are we, then, at the end point of this way of organizing nature? While avoiding (in fact, explicitly critiquing) the doomsday rhetoric found in much contemporary radical environmentalism, Moore believes we are now on the edge of an epochal crisis for capitalism. Rather than the apocalyptic "end of nature" that Bill McKibben identifies, Moore emphasizes an epochal end to "cheap nature."

In order to accept that capitalism is approaching this epochal crisis, we must accept the necessity of new frontiers of appropriation of cheap work/energy and ecological surplus, agree with his empirical analysis that "peak appropriation" has passed, and acknowledge that there are insufficient new frontiers to make nature cheap again.

We could express Moore's claim here in two forms. In the weak version, capitalism stagnates in sluggish growth without new frontiers (i.e., cheap natures) to appropriate. Expressed strongly, this means capitalism would cease to exist without

14. Moore, *Capitalism in the Web of Life*, 98.

cheap natures. We agree with the former but are not fully convinced of the latter, which seems closer to Moore's own position.

If we accept the first contention that frontiers are necessary, we must still consider the second: are they really exhausted? Is the era of cheap nature at an end? Here we seem closer to a question which could be answered empirically, and which for us is still open to debate. Moore's answer, and ours, to the question of how capital creates and absorbs frontiers depends on how we understand the nature-society relation.

For Marx, a dialectic is a mode for the presentation of categories; a way to reconstruct a complex totality of relations in thought.[15] Marx does not claim that reality *is* dialectical. But Moore collapses this distinction. This provides the basis for his somewhat quick rejection of "cyborgs, assemblages, networks, hybrids"—heuristics he apparently deems insufficient for theorizing the oikeios.[16]

Dialectics are concerned with internal relations: those relations that are essential to their terms. But not all relations are internal. This is why when Marx opens *Capital* with a discussion of commodities, he abstracts from their particular properties and the desires they satisfy, stating that whether "they spring from the stomach or from fancy, makes no difference."[17] Moore makes a similar distinction in discussing coal: "to paraphrase Marx, coal is coal. It becomes fossil fuel 'only in certain relations.'"[18] Typically, Marxists refer to this as a distinction between natural form (coal) and social form (fossil fuel), but in keeping with the spirit of the oikeios in Moore, a distinction between object and relational forms is more appropriate. This would also allow space for coal to occupy a wider range of relations with more than just its fuel-burning commodity role.

These relational categories can be reconstructed into a dialectical totality— tracing the web of connections whereby categories co-constitute each other. This is an important and powerful method. But Moore sometimes appears to commit a category error in dismissing a host of alternative ecological approaches (e.g., cyborg ecology). The result is a false antithesis in his analysis between Marxist-dialectical methods and cyborg networks. A messy cyborg ontology and a neat dialectical presentation of categories need not be mutually exclusive. And in fact, Moore's work *needs* the queerness of a cyborg communism. For example, while oikeios can

15. Bertell Ollman, *Dance of the Dialectic: Steps in Marx's Method* (Champaign: University of Illinois Press, 2003).
16. Moore, *Capitalism in the Web of Life*, 5.
17. Marx, *Capital* (1976), 125.
18. Moore, *Capitalism in the Web of Life*, 145, quoting Marx, *Wage-Labor and Capital* (New York: International Publishers, 1971).

serve as a general stand-in for relationality, it also has etymological resonances with *oikonomia*, the specifically familial (European, heteropatriarchal) understanding of the economy Angela Mitropoulos analyzes in *Contract and Contagion: From Biopolitics to Oikonomia* (2012). If left unaddressed, this resonance can fit quite easily with such an economic understanding of nature.[19]

Such a productive engagement requires restricting Moore's dialectic to its proper domain—a mode of presentation for internally related categories. The claim that reality is irreducibly dialectical, and hence all relations are internal, strikes us as untenable. It could become an obstacle to an ecopolitics that might turn science away from capital and the state by developing social relations for abstract scientific forms of knowledge to enter into connections with practical local knowledges.

Why does this split between nature and society persist? Although it is clearly necessary to capital's contemporary historical mode of organizing nature, Moore provides few insights into this ideological question. The late Marxist geographer Neil Smith argued the development of capitalism generated a contradictory ideology of nature as either a frontier to be conquered (capitalist modernity) or a pristine wilderness to be preserved (capitalist romanticism).[20] For Smith, nature/society dualism was the intellectual expression of real historical processes wherein frontiers really have been objectified as sources of raw materials and wilderness really has been created (such as by the clearance of Indigenous people to create national parks).

The bulk of Moore's analysis is avowedly geared towards assessing the situation capitalism has ushered in—and its proximity to possible collapse. While this is clearly important, the real test of such analyses is how they allow us to think through what is to be done politically in such a moment. Moore has surprisingly little to say about politics (aside from usual passing references to class struggle and an approving nod towards global food sovereignty movements). The main takeaway of the book is not to conceive nature and society as separate entities or objects and, instead, see them as historically produced and intertwined. But this is hardly a new insight; in fact, it is thousands of years old.

While Moore seeks to historicize capital's organization of nature through an analysis of successive energy regimes and agricultural revolutions, he misses the opportunity to historicize the nature/society dualism itself and thus to understand

19. See Mitropoulos, *Contract and Contagion*. On further dangers of full relationality, see Frédéric Neyrat, *The Unconstructable Earth: An Ecology of Separation* (New York: Fordham University Press, 2018).

20. Neil Smith, *Uneven Development: Nature, Capital, and the Production of Space* (Athens, GA: University of Georgia Press, 2008).

its persistence. "Nature" really does appear to capital as a series of frontiers to conquer; as resources and labor power to exploit; as a sink in which to dump pollution. This remains true even if, in fact, capital is a way of organizing nature and not an external force which encounters it. The nature/society dualism reflects capitalist modernity as it really appears: an ideology of nature.

Posed in its relation to nature, capitalism's current crisis can be reassessed as either developmental or epochal. To us, it will be epochal *only to the extent to which we participate in making it so*. Getting out of the ideology of Green Arithmetic requires much more than better thinking about or developing better language for the world we live in. It requires that we begin to operate as if nature were truly important to capitalism.

The political upshot of such a move is that our struggles against capital appear less symbolic and more material. They might be not quite as dialectical, but rather necessarily messy and antagonistic. In demonstrating them to be so, we show that struggle is not marginal to capitalism's demise but crucial to it. We need deeper and more coordinated global organization of ecological agitation, including blockades by workers, scientists, Indigenous peoples, farmers, and refugees. We need a planetary struggle. We cannot wait for capitalism's epochal crisis nor think our way into another world. We must begin building it today.

III

FUTURES

INTRODUCTION

TOWARD A REGENERATIVE UTOPIANISM

Several members of Out of the Woods found one another through involvement in the 2010–11 UK student movement. In that moment, Paul Mason's thesis of "the graduate without a future" mingled with the queer nihilism of Lee Edelman as slogans like "no future: utopia now!" and "work's shit, the future's shit, all I want is revenge" propagated through online networks and offline demos and occupations. This milieu became one of our departures for thinking about climate change.

On the face, it all seemed to fit. If tuition fees can cancel the future, then climate chaos certainly can. And indeed, mainstream environmentalism is full of appeals to think of the children, the coming generations whose futures climate change would seem to imperil. We reached for Edelman's critique of "reproductive futurism," the investment of futurity in the image of the child, and its rejection, a nihilistic embrace of "no future."

In some of our first essays, which we chose not to include in this collection, we attempted to read climate change through Edelman's theoretical lens. Shortly after we started down this path, however, we began to question it. Where Edelman embraces "no future," we insisted on appending "utopia now!" Edelman embraces nihilism, we rather identified "no future" with a ecological crisis that is here and now, having already rendered obsolete imagined futures that see climate change as something that looms menacingly on the horizon rather than as an unfolding, differentiated reality we're living inside. We want to stress the important roles that utopian imaginations can play, by making clear the contingency of the world in in which we live, and "educating our desire" such that we desire more and desire better.[1]

Given this understanding, we are a little more cautious: now it is not so much "no future" as *no to the future* as the continuation of a present that is impossible and undesirable. No to a politics that begins in the future rather than with an analysis of present tendencies. We must undo and remake the world. *In doing this we might return the future to us as a space of collective possibility.* In other words, by rejecting the future as the space around which our politics should be organized in the first instance we might actually pay the future the fidelity it deserves.

We've also become wary of the ways queer antinatalisms, such as Edelman's, can bleed into misogyny and hostility to pregnancy and kids themselves, while opening the door to Malthusian populationist tropes about "too many people." Edelman is careful to take aim at the ideological image of the child and not at actually existing children, but a politics which cares about those actually existing children and those who raise them—in addition to everyone else—requires more than a critique of reproductive futurism. Indeed, at a time when actual kids (and teens) are striking for climate justice and embracing left-wing politics, it particularly behooves us to think about young people as vital comrades in the struggle against climate crisis. They should not be fetishized, of course, just as to blame current adult generations for this crisis is to ignore contributions to and struggles against it. Youth need to be taken seriously as a political force. The tiresome old adage that people get progressively right-wing as they get older is unlikely to hold with a generation that, without dramatic change, will be unable to accrue the material benefits that might push them rightwards. Paradoxically, then, it is the understanding that there is, under present conditions, no future which is leading to the possibility, once again, of *a future* in the most profound sense as that which is radically, unknowably different from the present.

What this might allow us to think about is a *regenerative utopianism*. In contrast to reproductive futurism's hope that "our" children will enable the future to unfurl as a continuation of the (white supremacist, colonial, heteropatriarchal, capitalist) present, regenerative utopianism sees struggles around the raising of children and against the biological family as vital terrains from which the future might

1. See David M. Bell, *Rethinking Utopia: Place, Power, Affect* (London: Routledge, 2017).

be opened up to possibility.[2] These futures might be theorized with readings of utopian fiction such as Marge Piercy's *Woman on the Edge of Time*, though it would read these texts as heuristics meant to catalyze or orient change and not as blueprints to be realized. It might also draw on the expanded notions of kinship found, for example, in Citizen Potawatomi Nation member and environmental biologist Robin Kimmerer's *Braiding Sweetgrass* (2014). For Kimmerer, to be Indigenous means to take care of the ecologies of place such that future generations of human and nonhuman kin might be able to use them for education, food, medicine, and enjoyment.

In "The Future is Kids' Stuff," we engage with Edelman's critique of reproductive futurism. OOTW member Sophie Lewis' "Cthulhu Plays No Role for Me" takes up a similar theme from a different starting point. While written in an individual capacity—indeed, it forms a deeply personal falling-out-of-love letter with cyborg-philosopher Donna Haraway—it also marks our collective shift from uneasy deployment of queer antinatalism to something of a queer a-natalism. Here the critique of reproductive futurism mutates into a decentering rather than a repudiation of reproductive sex, one which rejects the injunction to "make kin not babies" and its associated populationist politics for a queer politics with a place for kids and kin-making alike.

The final piece in this section is written in response to the publication of Nick Srnicek and Alex Williams' *Inventing the Future* (2015). OOTW member Joseph Kay's "Postcapitalist Ecology" discusses how the image of nature used in theorizing the politics of the future shapes the resource streams accounted for by technoutopian futures. In particular, leaning too hard on technological transformations threatens to leave neocolonial divisions of toxic extraction and waste intact. If modernity is not to be a synonym for the coloniality of power, the imbrication of the two cannot be disavowed.

---

2. Sophie Lewis, *Full Surrogacy Now: Feminism Against Family* (London and New York: Verso Books, 2019).

THE FUTURE IS KIDS' STUFF

First published May 2015

Why does the figure of the child play such a central role in motivating action against climate change, and with what consequences? How does such "reproductive futurism" foreclose certain queerer forms of futurity, or the concept of 'the future' itself? This essay was an attempt to think through such questions in the wake of reading one somewhat troubling example in the conclusion to Naomi Klein's This Changes Everything *(see "Blockadia and Capitalism" in this volume). It was also an attempt to come to terms with and supersede our prior use of "no future" as something of an Out of the Woods slogan, a phrase we now reject. At stake instead should be the question of which future and which forms of human and nonhuman kinship and comradery are at stake in different pasts and futures.*

"Won't somebody please think of the children?" This is the pithy question espoused by policy makers, media platforms, and environmental NGOs. The catchphrase is meant to make the consequences of climate change meaningful to the most be- loved subject of liberal political economy: the family. We are urged to give a shit about climate change and consequently modify our lives (by driving less, turning our lights off, voting for a particular party, or marching from A to B on a demo) because the futures of our descendants are dependent upon the environmental conditions we create today. The innocent face of the child looming in the future demands that we enroll ourselves in a particular political field and, thus, in a particular form of politics.

We have previously argued against the future promise of environmentalism, embracing the slogan of punks and queers: no future.

> As an utter necessity we must abandon the future, for we cannot win there. No future, for we will never convince the majority to fight for the sake of a time they cannot imagine. No future, for capital will always defeat any strategy based on a next-ness, for against airy notions of tomorrow's world, they can posit the cold hard facts of today counted out in wages and jobs. No future, because, right now, there is literally no future, right now we are condemned to collapse.[1]

Nearly everywhere we look, we are confronted by forces that call us to action in the name of that which is yet to come. But the demand of future human generations, which Lee Edelman hyperbolically calls the "fascism of the baby's face," constitutes an overwhelming and overarching logic that structures the symbolic field of nearly all politics.[2] In the regime of this pervasive reproductive futurism, we are taught to ignore the demands for and enactments of "utopia now." We deprioritize demands emanating from our contemporary catastrophe, urgent irruptive calls for altered modes of life and of living together that begin in and push against the present. In submitting to the faith that tomorrow will be rendered hospitable for the child, we defer our responsibilities to the present (and actually existing children, for that matter, who contribute much to what is good about our present).

---

1. Out of the Woods, "No Future," The Occupied Times (blog), October 22, 2014, https://theoccupiedtimes. org/?p=13483. See the introduction to this section for a more nuanced take on our contemporary orientation to the future. This continues to reject the future as the de facto form to which climate politics should be oriented, but argues that by struggling in and against the present we might open up the future as a space of possibility once more. It is less a case of dealing with a future that is yet to come and more of fighting for a future we would like to come. This essay has been modified to more closely reflect this position.
2. Edelman, *No Future*, 75.

These politics of the child are structurally conservative. Children are figured as the unit through which private property is passed and through which gendered, racialized, and classed society is reproduced. Nature, meanwhile, is rhetorically constructed as both "out there" spatially (we are urged to reduce our ecological "footprints") and "out there" temporally (the threat always looms on the horizon; it's always our "last chance" to save the children). What is excluded are the conditions of social reproduction, the multiplicity of forms of kinship, and the immanent possibility of rupture. Against the structurally conservative politics of reproductive futurism, we need the structurally open politics of regenerative utopianism.

REPRODUCTIVE FUTURISM

Reproductive futurism saturates the political to its core. As Edelman argues, political futures are rendered so dependent on this symbolic child (who is quite obviously heteronormative but also white, suburban, etc.) that our politics ends up fighting over the image of the child rather than cutting across its fictions. Edelman argues, "The image of the Child, not to be confused with the lived experiences of any historical children, serves to regulate political discourse—to prescribe what will count as political discourse—by compelling such discourse to accede in advance to the reality of a collective future whose figurative status we are never permitted to acknowledge or address."[3] That is, reproductive futurism demands the repetition of the status quo and all its violences.

By contrast, Edelman shows, the figure of the queer is rendered a threat to the symbolic and moral order. Homosexuality and nonconformity with cis-sexist gender norms are positioned as affronts to the family and thus to life itself. It is not difficult to find and criticize examples of "family values" rhetoric coming from the far right (and the transphobic liberals who so frequently intersect with them), but through the lens of reproductive futurism this structurally queerphobic element becomes visible in a far broader swathe of mainstream politics. One can easily begin to see the ways discourses of the Left, and particularly the Green Left, are complicit with this queerphobia through their overt reliance on figures of reproductive futurism.

Once we understand reproductive futurism, we begin to find images of the child and of actual children in liberal ecological politics. The climate science documentary *Thin Ice* opens simply and strikingly, with a silent, extreme

3. Edelman, *No Future*, 11.

close-up of a white infant's pursed lips and blue eyes. In the run-up to the fifteenth Conference of the Parties (COP 15) at the 2009 United Nations Climate Change Conference—a global summit on climate change typically held in Copenhagen, Denmark—billboards enjoined their audiences to "Hopenhagen" for the sake of the futures of small blonde children. A recent protest of LEGO's alliance with Shell involved fifty children "building their favorite Arctic animals out of oversized LEGO bricks," linking both the futurity of children to that of nonhuman species.[4]

Despite dwelling occasionally on the possibility of a different, queerer, partially nonreproductive politics, even Naomi Klein holds up the naturalness and desirability of birthing and reproduction as a beacon.[5] At stake in these discourses is always the defense of an imagined proper, natural, and secure social order. Edelman argues, "Nature [is] the rhetorical effect of an effort to appropriate the 'natural' for the ends of the state. It is produced, that is, in the service of a statist ideology that operates by installing pro-procreative prejudice as the form through which desiring subjects assume a stake in a future that always pertains, in the end, to the state, not to them."[6] Nature thus plays a key role in guaranteeing the persistence of reproductive futurism.

To highlight the pernicious effects of reproductive futurism is not to be against children, reproduction, or heterosexuality as an orientation per se. Rather, it is to show that when reproductive futurism structures politics we become incapable of developing forms of political thought and practice demanded by ecological crisis. Reproductive futurism denies agency to children, rendering them mere vehicles for our political desires. Without putting in place material conditions of reproductive justice for women, it celebrates women's capacity to gestate and birth infants as a free service for capital. It ropes queers into the futile endeavor of proving themselves to be "properly" worthy white settler citizens with supposed family values.[7]

---

4. Camilla Møhring Reestorff, "'LEGO: Everything Is NOT Awesome!': A Conversation About Mediatized Activism, Greenpeace, Lego, and Shell," *Conjunctions: Transdisciplinary Journal of Cultural Participation* 2, no. 1 (2015): 30.

5. See "Blockadia and Capitalism" in Section IV of this book.

6. Edelman, *No Future*, 52.

7. Commenting on the increasing acceptance of "homonormative" lesbian and gay families, Jasbir Puar argues that that "the capitalist reproductive economy (in conjunction with technology: in vitro, sperm banks, cloning, sex selection, genetic testing) no longer exclusively demands heteronormativity as an absolute; its simulation may do." [Puar, *Terrorist Assemblages*, 31.]

Reproductive futurism not only structures the heteronormative desires of the political field and social order, it also renders racialized, Indigenous, and migrant peoples as threats to that order. In doing so, it affirms the management, disposal or murder of such peoples by the state and capital. In "Decolonizing Feminism: Challenging Connections between Settler Colonialism and Heteropatriarchy" (2013), Maile Arvin, Eve Tuck, and Angie Morrill compel us to take into account "the pervasive violence perpetuated on Indigenous peoples through campaigns focused on managing Indigenous reproduction and child-rearing (from boarding schools to eugenics and forced sterilization)." In the face of such tactics, they argue, "proposing to invest in 'no future' seems not only irrelevant to Indigenous peoples, but a rehashing of previous settler colonial tactics."[8]

A further instance of this logic is at work in Palestine. Israeli settler colonialism is typified by visceral and arbitrary violence which attempts to produce just such a state of continually endangered existence outside the recognized social order. Eyal Weizman notes the existence of an Israeli military document called "Red Lines," which specifies the precise threshold quantities of vegetables, meat, milk, and other "essentials" below which mass starvation—and charges of genocide—would result. The Israeli state then seeks to restrict aid flows into the Gaza Strip to this threshold level. Weizman quotes Dov Weisglass, an advisor to former Israeli Prime Minister Ehud Olmert, saying: "the idea is to put the Palestinians on a diet, but not to make them die of hunger."[9]

The Palestinian population is thus forced to bear a survival circumscribed by its proximity to death, a situation Achille Mbembe calls "necropolitical."[10] Power is played out through a combined destruction of social and human life. To put a people so close to death is to position murder as a minor escalation or even an accident. The genocidal tendency that animates the occupation emanates from this proximity.

What does reproduction mean in such a context? In July 2014, Israeli lawmaker and member of parliament Ayelet Shaked wrote on Facebook that the Palestinian people

---

8. Maile Arvin, Eve Tuck, and Angie Morrill, "Decolonizing Feminism: Challenging Connections between Settler Colonialism and Heteropatriarchy," *Feminist Formations* 25, no. 1 (2013): 24.

9. Eyal Weizman, *The Least of All Possible Evils: Humanitarian Violence from Arendt to Gaza* (London and New York: Verso Books, 2011), 83. See also Jasbir K. Puar, *The Right to Maim: Debility, Capacity, Disability* (Durham: Duke University Press, 2017).

10. Mbembe, "Necropolitics."

. . . are all enemy combatants, and their blood shall be on all their heads. Now this also includes the mothers of the martyrs, who send them to hell with flowers and kisses. They should follow their sons, nothing would be more just. They should go, as should the physical homes in which they raised the snakes. Otherwise, more little snakes will be raised there.[11]

The next day, three Israelis snatched Mohammed Abu Khdeir, a sixteen-year-old Palestinian, from a street in East Jerusalem. The police would later find his body in a nearby wooded area. Shaked was appointed as justice minister a year later. Here we see clearly that reproductive futurism is not a monolithic entity pertaining to all children, but rather a series of antagonistic and mutually exclusive articulations. The occupation puts the Palestinian child into fundamental conflict with the children of Israel. Indeed, whenever reproductive futurism is enrolled by the nation-state, delineations of which children's lives are rendered livable or expendable become visible. To Israel, the future of its children is fundamentally threatened by the presence of the other-kid, the "little snakes" who present a mortal danger to their young ones.

In this, it continues a long line of settler colonial behavior practiced around the world. The US Army colonel John Chivington justified the slaughter of Indigenous children of a number of nations in what has become known as the Sand Creek Massacre of 1864 with the oft-repeated phrase, "kill 'em all, big and small, nits make lice." As Katie Kane has noted, this phrase can be traced further back in colonial history, being used by an anonymous English poet in 1675 to commemorate the massacring of the residents of the Irish town of Drogheda by Oliver Cromwell's men under the command of Charles Coote, who "[d]id kill the Nitts, that they might not growe Lice."[12] Noting that this use of "louse" positions infant colonized subjects as future threats to the colonial order, Kane also invokes Frantz Fanon's metaphoric reference to DDT, which "destroys parasites, carriers of disease, on the same level as Christianity, which roots out heresy, natural impulses, and evil."[13]

In the context of the antagonism of national futures, the Indigenous or colonized child exists in contradiction to the children of the settler. If "no future" were to merely mean that one "not fight for the children," one would abdicate responsibility for the fact that children (and images of the child) are often fighting to

---

11. Ali Abunimah, "Israeli Lawmaker's Call for Genocide of Palestinians Gets Thousands of Facebook Likes," *The Electronic Intifada* (blog), July 7, 2014, https://electronicintifada.net/blogs/ali-abunimah/israeli-lawmakers-call-genocide-palestinians-gets-thousands-facebook-likes.

12. Quoted in Katie Kane, "Nits Make Lice: Drogheda, Sand Creek, and the Poetics of Colonial Extermination," *Cultural Critique*, No. 42 (Spring 1999): 84.

13. Fanon, *The Wretched of the Earth*, 7.

destroy each other. This nationalistic logic is clear in the xenophobic approach to refugees, with the European Union proposing joint military action to destroy boats in order to prevent the flow of refugees fleeing Libya.

Therefore, there is not one overarching symbolic child, but rather a multitude of antagonistic children that reproductive futurism—especially in global climate change discourse—deftly papers over in the name of the unmarked child-as-such. The political system ultimately recognizes the need for children within society, yet at the same time it has very little desire to aid actually existing children and their kin.

Reproductive futurism, then, is a vital component to an understanding of capitalist heteropatriarchy but it is not a universal structuring field of politics as Edelman suggests. Indeed, this lack of nuance means that we must be wary of merely inverting the binary opposition of "pro-reproduction" and "anti-reproduction" without challenging the very grid of intelligibility which centers reproduction in the first place.[14] To note the prevalence of reproductive futurism in climate change politics invites us not simply to revalue nonreproduction or anti-futurity, but to go beyond the political limits of centering reproduction and of thinking the time yet to come through reproduction.

KINSHIP OF THE INFERTILE

Naomi Klein gestures towards the possibility of a "kinship of the infertile," a regenerative politics that would allow us to escape reproductive futurism. While reproductive futurism infects the vast mass of climate change politics, there is also within that mass the promise of a new, utopian (rather than simply futuristic) movement that shifts itself towards the politics of what we would recognize as regenerative cyborgs.

Like Donna Haraway, we would rather recognize our finitude and our partial and everyday situations within networks of technology, care, and communication than within the imperial and destitute politics of all future children. For Haraway, the cyborg is "resolutely committed to partiality, irony, intimacy, and perversity. It is oppositional, utopian, and completely without innocence. No longer structured by the polarity of public and private, the cyborg defines a technological polis based partly on a revolution of social relations in the *oikos*, the household."[15]

14. "By assuming that reproduction is at the center of futurity and the platform against which future-negating queer politics should be orientated, Edelman . . . ironically recenters the very child-privileging, future-orientated politics he seeks to refuse." [Puar, *Terrorist Assemblages*, 211.]

15. Haraway, *Manifestly Haraway*, 9.

The politics of regenerative cyborgs would thus refuse the pure and innocent future imagined and defended by reproductive futurism, instead understanding our present selves and our lives with children and other species as transversed by technical, social, and queer connections. Instead of focusing on the ephemeral images of (white) children, it would instead reorient us towards the material conditions of our lives. This would, as Silvia Federici argues, reopen "a collective struggle over reproduction, reclaiming control over the material conditions of our reproduction and creating new forms of cooperation around this work outside of the logic of capital and the market."[16]

By broadening the grounds of the struggle over reproduction to include the general material conditions of social life, the politics of regenerative cyborgs would take into consideration the agriculture we will need to feed ten billion people, the oceanic gyres, jet-streams, and weather systems that will shape lives and the carbon cycle which itself shapes those forces. To posit ourselves as regenerative cyborgs pushes us to think of our partial positions within a complex network composed of machines and organisms, as an entanglement of living and nonliving things on cyborg Earth.

In this vast web, we cannot reduce regeneration to mere reproduction, with its tight bonds to heteronormative and survival-based notions of human life. What is being reproduced is far greater than the merely human. This is why, what is more fitting here is a politics of regeneration, such as the one as described by Jasbir Puar: "what is at stake [is] not the ability to reproduce, but the capacity to regenerate, the terms of which are found in all sorts of registers beyond heteronormative reproduction."[17] Through regeneration of the material social infrastructures of care that makes our lives livable and meaningful, we begin to imagine the transformative gesture of our previous call for "no future." No future means nothing without its corollary, "utopia now." This demand stems from our desire and capacity to regenerate ourselves in ways falsely held to be impossible: to expand these capacities from the cramped spaces in which they currently develop into an unknowable beyond where we might live lives of complexity and beauty, free from the horrors and constraints of ecocidal, white supremacist, capitalist heteropatriarchy.

Rather than no future being an alternative to reproductive futurism, we can therefore see it as a reflection of it under conditions of ecological crisis. In the context of business-as-usual climate change, the reproduction of the same is no

16. Federici, *Revolution at Point Zero*, 111.
17. Puar, *Terrorist Assemblages*, 211.

future. A politics of regenerative cyborgs brings the conditions of social reproduction back into view, thus opening the possibility of the nonreproduction of the same: utopia now.

We can regenerate and emerge as the cyborgs we always already are. A cyborg praxis gives us the space to understand difference, to see that reproductive futurism desires little white settlers for children, and seeks to destroy all those who are young and outside of this category. A cyborg praxis allows us to see that the fetishism of the child in the abstract is inseparable from the actual and total violence perpetrated against children and their kin. But perhaps more than anything else, a cyborg praxis offers us a chance amidst the ruins. It means the refusal of the lifeboat ethics demanded by the defense of the national child, but also the embrace of that which lies beyond them: the entwined possibilities of regeneration and transformation.

CTHULHU PLAYS NO ROLE FOR ME

First published May 2017

Donna Haraway is a crucial thinker for many of us; it is with reference to Haraway that we frequently think of OOTW as an advocate of queer cyborg communism. Thus, it was with considerable disappointment that we found in the pages of 2016's Staying with the Trouble: Making Kin in the Chthulucene *a certain amount of underlying 'populationist' thinking, shockingly reminiscent of Garrett Hardin's "lifeboat ethics," described in our essay "The Dangers of Reactionary Ecology." In this essay, OOTW member Sophie Lewis describes their (and our collective) affinity with Haraway's earlier thought and the implications and consequences of her turn towards concern with overpopulation and populationist thinking.*

Donna Haraway's most famous piece of writing declared itself "faithful to feminism, socialism, and materialism."[1] But in the eighties, there were many feminists, socialists, and materialists who couldn't see how this self-described "political myth" was faithful to them at all. Was comrade Haraway recommending—beyond even critically embracing technology, as the Bolsheviks had—incorporating it into the human body? In fact, Haraway's "A Cyborg Manifesto" (1985) expressed a dream of a politics neither of repudiation nor exodus but rather—as she put it—faithful irony (i.e., blasphemy) vis-à-vis heteropatriarchal, racial technocapitalism. She encourages something like a killing embrace of the brutal either/ors and deadly dyads imposed by that "matrix" of power onto would-be human subjects. She invites recognition of one's individual (uneven) imbrication with colonialism and the military-industrial complex—the better to fuel one's loving rage and fervor to dismantle those evils. As a trained biologist and primatologist, Haraway is able to deliver her blasphemy in a formally blasphemous blend of scientific and poetic tongues.

Haraway's multiple "cyborg" articulation of the self as a kind of proletarian drag proved to have intense resonance across the world. It is, in its own words, "oppositional, utopian, and completely without innocence."[2] If the success that greeted the manifesto surprised its author, the suspicion and shock did not. A significant legacy of the antinuclear and antimilitary organizing throughout the long seventies—the feud in which, in fact, Haraway was intervening—consisted of a false antithesis between a convinced technophobic leftism and practically all other approaches to the matter of *techne*. Of course, at the time, myriad Marxist writings on the imbrication of capitalist technologies with natural entities existed, which reflect on the possibility of salvaging them for emancipatory ends—including the account of "nature as an accumulation strategy" developed by eco-Marxists Cindi Katz and Neil Smith. But few such interventions were coming (ostensibly) from within eco-feminism, the camp of peace activists and Earth's defenders. Haraway, the self-described "Sputnik Catholic," did belong, in part, to that camp.[3] Nevertheless, her sisters in grassroots feminism proved reluctant to take her message on board. The cyborg's popularity surged elsewhere, notably in art and urban studies. David Harvey hailed her as an indispensable figure for the practice and study of spatiality: "she has evolved a wonderful way of talking that acknowledges that, if everything

1. Haraway, *Manifestly Haraway*, 5.
2. Haraway, *Manifestly Haraway*, 9.
3. Haraway, *Manifestly Haraway*, 283.

is related to everything else in the world, then we must create sentences to reflect that fact."[4]

Thus, while "A Cyborg Manifesto" was originally meant to be a straightforward contribution to the erstwhile left publication *Socialist Review*, what emerged was gobsmackingly "postmodern." As Haraway boldly declares, "The dichotomies between mind and body, animal and human, organism and machine, public and private, nature and culture, men and women, primitive and civilized are all in question. . . . they have been 'techno-digested.'"[5] Rephrasing Bruno Latour's famous dictum "we have never been modern," she advanced that "we have never been human."[6] In a 1996 interview, she clarified this problem in more conversational terms: "Thinking of machines as an 'it,' over and against which our organic and internal cells have to conduct some kind of heroic struggle, is a very hard framework to avoid."[7] Even as it bewildered and offended elements of the Left—who declined to see why one might want to avoid that kind of framework—the manifesto offered nourishment to many others. The new analytic weapons it proffered invited mutant, contaminated subjects to build a new world on the ruins of the present-day home, factory or lab. "Cyborg," for some of us, is a luminous translation of the Marxist idea that we make history but not under conditions of our choosing.

Yet, for all its polemical antihumanist pizzazz, cyborgicity was grounded solidly in social reproduction theory of the kind pioneered by Marxist feminists Nancy Hartsock, Ruth Cowan, and Barbara Ehrenreich. What separates Haraway's manifesto from their meticulous dissections of labor divisions and market transformations is, rather, its seemingly miraculous syncretism. Black, Indigenous and Latinx feminisms (e.g., bell hooks, Audre Lorde, Barbara Smith, Cherríe Moraga, and Gloria Anzaldua), lesbian and "deconstructive" feminisms (e.g., Monique Wittig), and queer, anticolonial afrofuturisms (e.g., Octavia Butler) were all treated as though they were always already inextricably linked to conversations in biology about genes, computer chips, symbiogenesis, and cybernetic matrices (in particular the critiques of science of Sandra Harding, Richard Lewontin, Hilary Rose, Zoe Sofoulis, Stephen Jay Gould, et al.).

"A Cyborg Manifesto" is, in some ways, a retelling rather than reinvention of emancipatory thought's fundamental "eco vs. techno" dialectic. At the time of

4. David Harvey and Donna Haraway, "Nature, Politics, and Possibilities: A Debate and Discussion with David Harvey and Donna Haraway," *Environment and Planning D: Society and Space* 13, no. 5 (1995): 508.
5. Haraway, *Manifestly Haraway*, 32.
6. Nicholas Gane, "When We Have Never Been Human, What Is to Be Done? Interview with Donna Haraway," *Theory, Culture & Society* 23, no. 7–8 (2006): 135–58.
7. Harvey and Haraway, "Nature, Politics, and Possibilities," 514.

its publication, contradictions in this arena were coming to a head within feminism over the "new reproductive technologies." The essay combines economic analysis—of the centrality of "homework" to the tech "revolution"—with a deconstruction of the figure of the "human" in many ways reminiscent of Frantz Fanon's. Like Fanon, Haraway centers the queer and racialized character of the animal "proletarian." Applying her cyberfeminist primatologist's eye, she also insists "we are all chimeras," historically situated implosions of animal and technology, virtuality and physicality.[8]

If we have never been human, then what have we been? What are cyborgs? Part of the answer, to many readers' surprise, is—simply—"women." In calling up this "invisible," "leaky" virtual monster, she calls on a mass constituency to recognize and reimagine itself.[9] "Women and other present-tense, illegitimate cyborgs, not of Woman born, who refuse the ideological resources of victimization so as to have a real life."[10] In the same way that Wages for Housework was a weaponization of wages against housework, however, the invocation of women was intended to "abolish gender." Haraway held forth "a picture of possible unity . . . the self [that] feminists [of all genders] must code" so as to foment a state of being "responsible" to the social relations of science and technology in all their contingency.[11] Politics, she suggested, "means both building and destroying machines, identities, categories, relationships, space stories."[12] Reconstructing the boundaries of daily life would inevitably yield a new human-ish subject "in partial connection with others, in communication with all of our parts."[13] The machine and the monster, she explained, are both "us."[14]

By the time I encountered "A Cyborg Manifesto" in the early 2000s, it was a cult text. Haraway penned her 1985 chimeric essay in California, a context that inflects its part binary-imploding "fabulation" of liberatory subjectivity, part avant-garde account of the political economy of post-Fordist societies' production of people.[15] Somehow, to me, Harawayian words felt like coming home. Her writing was witchy and baroque, yet easy to read. But I soon learned that

8. Haraway, *Manifestly Haraway*, 7.

9. Haraway, *Manifestly Haraway*, 13 and 11.

10. Haraway, *Manifestly Haraway*, 59.

11. Haraway, *Manifestly Haraway*, 30 and 57.

12. Haraway, *Manifestly Haraway*, 68.

13. Haraway, *Manifestly Haraway*, 67.

14. Haraway, *Manifestly Haraway*, 65.

15. Her assignment had been to map the contours of socialist feminism in the era of neoliberalism and the Web. The East Coast *SR* editorial collective hated what she turned in, but the West Coast collective overruled them and published it (and thank goodness they did).

many cultural gatekeepers in my British environment—notably those wedded to George Orwell's patronizing and spartan rules for avoiding English pretentiousness—simply couldn't stand it. Such people insisted that, "objectively," these mad, dense sentences of hers just weren't clear. Their displays of nonplussed intolerance in the face of Haraway's rococo prose seemed suspiciously disproportionate. Might they simply represent the cost of being a "FemaleMan" treading cheerfully and irreverently on Marxological ground?

I think so, but I am anything but impartial. After all, Haraway began as my hero. My comrades teased me relentlessly for citing her in every single one of my articles and reaching for her in every conversation. Admittedly, my interest was mainly in the older stuff, like *Modest_Witness@Second_Millenium.FemaleMan©_Meets_OncoMouse™: Feminism and Technoscience* (1997), where Haraway criticizes the "obsession with the gene as a form of reification similar to that of commodity fetishism."[16] By the time I discovered her writings on the cyborg, she was talking about dogs (a political choice, she claims, but not one whose implications were ever clear). At that point, she overwhelmingly had, in Alyssa Battistoni's words, "more to say about kin-making and agility competitions" than political coalitions and oppositional strategies.[17] Still, it was often fascinating stuff, posing questions like: what kind of world do nonhuman beings want? How can we know? I sat tight. I had, after all, felt firsthand her ability to make Marx readable, relevant, and cyborg for polymorphously perverse teenaged girls like me. It is thanks to her I came to anticapitalist thought and struggle in the first place. Were it not for Haraway, I might never have dared or desired to read Marx at all.

Here was a trained biologist who analyzed the swarming web of earthly life at the cellular level and pursued a revolutionary's desire for liberation in the same breath. She projected an infectious, subversive sense of confidence that biological realities—from computers to embryos to brains—militate for, rather than against, a comradely and just existence. Her thought laughs generously at humanism and posthumanism in equal measure, revealing playful and at the same time utterly serious modes of organizing, or lines of flight, within the deadly matrices of technology-mediated violence she insists our own bodies co-compose. Her unbounded, psychedelic, militant-particularist materialism doesn't so much explode the reproduction/production distinction as make it look ridiculous and embarrassing. She seems to be grimacing in the face of such categories, following

---

16. Alyssa Battistoni, personal correspondence, 2017.

17. Alyssa Battistoni, "Monstrous, Duplicated, Potent," *n+1*, April 12, 2017, https://nplusonemag.com/issue-28/reviews/monstrous-duplicated-potent/.

Medusa—a practice of reflecting back and splintering chauvinist epistemes which she calls "diffraction."

The figure of the cyborg turns the marginality of "queer" on its head, taking for granted that proletarian monsters under fire from transphobia and antiblackness are powerful recombinant operatives, central to class struggle. Rather than adding axes of oppression to her militant Marxian heuristic, she composted them. Her mission? To implement forms of organizing capable of uniting "witches, engineers, elders, perverts, Christians, mothers, and Leninists long enough to disarm the state."[18] She articulates a materialism that makes palpable how we are all touched by the cyborg virus in the "feminizing" landscape of neoliberal work. Though a story about common ground, it is not a sexy story. As Battistoni remarks in her own portrait of Haraway: "The Manifesto's popularity has no doubt been fueled in part by the vision of a bionic babe implied by the word itself—a Furiosa or the Terminator—but little could have been further from her meaning."[19] Battistoni's essay reminds its readers how, when asked to give an example of what exactly cyborgicity is in an interview, Haraway talked about "how like a leaf I am," describing the "intricacy, interest, pleasure, and intensity" of this sense of imagined kinship. How many people in 1989 (or since) pictured the neoproletariat as . . . leafy? Yet it is an intimate mass network of synthesizers, imperfectly communicating, individually mutating, and crackling with static.

At the same time, the image was less a question of acknowledging overlapping DNA, as Battistoni says, than "thinking about the immense amount of labor and practice that had gone into producing the knowledge that she was like a leaf in so many ways. Thinking about how incredible it was to be able to know such a thing."[20] Cyborgs are collective brains.

> Some folks pick up the figure of the cyborg and use it in a celebrational mode, and miss the argument that the cyborg issues specifically from the militarized, indeed a permanently war-state based, industrial capitalism of World War 2 and the Cold War. They miss that the cyborg is born as the cyborg enemy. . . . Now, from that particularly unpromising position, what possible kinds of cracks in the system of domination could one imagine beyond a kind of sublimity, a kind of wallowing in the sublime of domination which, of course, many folks do.[21]

18. Haraway, *Manifestly Haraway*, 16.
19. Battistoni, "Monstrous, Duplicated, Potent."
20. Battistoni, "Monstrous, Duplicated, Potent."
21. Harvey and Haraway, "Nature, Politics, and Possibilities," 514.

As Haraway's concern makes clear, far from representing an aestheticized apocalyptic ideal for the Anthropocene, the cyborg is a multiply colonized test-subject, situated squarely in the Capitalocene. "She" is a laborer who traffics in information, capital, and gender codes. Think of a hormone-deprived prisoner; or a manufacturer of low-grade circuit boards and computer chips working the night shift; or a pregnant housewife who is also a call-center contractor; or a forcibly sterilized migrant hijacking radio waves, evading searchlights.

Or, for instance, me. Ever since she first hacked my teenaged frontal lobes, I have made sense of myself as cyborg and stalwartly defended what I recognized in my marrow to be the funny, the wild, the profound, the radically illuminating genius of Haraway. I've argued against all of the standard charges laid against her: self-indulgence, stylistic obscurantism, "postmodern" triviality, etymological shamanism.[22] And, since Haraway opened the door to radical thought for me, what follows has been painful to write. It is a lamentation: not that her critics were right before, but that, substantively, her latest monograph, *Staying with the Trouble: Making Kin in the Chthulucene*, jumps the shark and heralds a change.

Since the eighties, a steady succession of characters have cropped up in Haraway's thought who are comparable figures to the cyborg but far less popular, and far less politically generative: the "modest witness", the coyote, the trickster, FemaleMan, the Surrogate, the companion species, Oncomouse™, string figures, and chthonic ones. Already in 2003, she declared in disgruntled tones that: "I have come to see cyborgs as junior siblings in the much bigger, queer family of companion species."[23] Indeed, in retrospect, I wonder now whether the coining of those more overtly "organic" successors must be understood in the context of Haraway's frustration with the persistent revolutionary humanism of cyborgicity. Perhaps I am sensing a double frustration in Haraway: not only with the common misunderstanding of the cyborg as a kind of android, but also with the very non-misunderstanding of her cyborg. Perhaps this cyborg, which Haraway called the illegitimate offspring of militarism, capitalism and state socialism, also represents a somewhat illegitimate (even partially regretted) daughter of a reluctant intellect.

In *Staying with the Trouble*, Haraway has made a decisive turn towards a primitivism-tinged, misanthropic populationism. Though she started off championing

---

22. I thought I could sometimes detect the whiff of misogyny on them anyway—the structure of thought that recoils subconsciously and willfully hears only babbling narcissism because it dimly registers a threat or feels indignant at not being the intended interlocutor.

23. Haraway, *Manifestly Haraway*, 103.

the cyborgs of class struggle against the goddesses of technophobia (her immortal closing line: "I would rather be a cyborg than a goddess"), my suspicion is that, now, she's gone over to the goddesses.

Despite enduring decades of denigration from some quarters on the Left as a "po-mo" thinker, Haraway's remarks about Marxism's limitations in the past have not remotely amounted to anti-Marxism. Haraway's "Cyborg Manifesto" cared deeply about human people in all their proliferating ingloriousness and it desperately wanted postgender communism for us—the species that reads and writes manifestos. It didn't link laboring with healthfulness, morality or being deserving. But in the essays constituting *Staying with the Trouble* she seems to have developed a new affinity for just that. Now, she wants a decline in human beings more than she wants to smash capitalism. In fact, it isn't clear if she even still wants the latter. Although the lines are drawn coyly, they are unmistakable. Her cursory but emphatic and repeated antinatalist instructions against making babies seriously risk rehabilitating the very eugenic antihumanism her early work on "Teddy Bear Patriarchy" inveighed against so brilliantly. Population reduction, as she now fantasizes it, is declared by fiat to be nondiscriminatory, friendly, collective, and noncoercive.

One would be justified in expecting to get some elaboration on how the removal of eight billion heads from the total headcount over the next century or so could be noncoercive—indeed, non-genocidal. But there is really only a fable, based around a micro-community in the United States, proclaiming that this is possible.[24] The utopia of two to three billion human beings is supposed to arise from a choice, simply, to not make babies. As a program, this represents a provocative break with materialism. It is also a provocation impossible to ignore or overlook, since it is effectively all that ties together what would otherwise be an unconvincing but inoffensive collection of vague, repetitive chapters on various eco-techno-animalian assemblages such as carrier pigeons and pills that stop urinary leakage in mammals.

The trend already seemed apparent in her last book *When Species Meet*, but it has now been consolidated, where the feminist-scientific emphasis on epistemic partiality (pioneered since the eighties by Haraway alongside figures like Sandra Harding) has turned into a commitment to pluralism, and where she actively shuns the pursuit of systems theorizing—for, as she says in a recent biopic, such theorizing

24. "The Camille Stories," as Haraway explains in *Staying with the Trouble*, are the upshot of a group science-fiction exercise in which Haraway participated with Vinciane Despret and Fabrizio Terranova at a summer workshop organized by Isabelle Stengers in France.

only ends up "dazzling" us.[25] Haraway's former (profoundly system-oriented) Marxian technofeminism has given way, then, to something called multispecies feminism. This tendency, pioneered also by Anna Tsing, seems to be characterized by a barely disavowed willingness to see whole cities and cultures wiped from the planet for the sake of a form of thriving among "companion species" involving relatively few of us.

RESPONSIBILITY FOR POPULATIONS

The revolutionary spark in Haraway's "more-than-humanism" has apparently been lost along the way and supplanted by an apolitical notion of trans-species *Gemeinschaft* [community]. One intriguing consequence of the place humanity now occupies in her ends-means argument is that nonhuman as well as human animals receive less methodological care. Even as Haraway's delightful critiques of Derrida and Deleuze's inability to really respond to actual animals (a cat, a wolf . . .) ring fresh in my ears, it strikes me that these "chthonic ones" (the latest case studies) are oddly distant and inanimate sketches of butterflies, spiders, pigs, ants, sheep, and racing pigeons. They are all "critters" by whom Haraway says (frequently) she is entranced, riveted, passionately gripped. Yet I don't see it.

The failure to respond to earthly companions, of course, is the very thing Haraway always sets out to consequentialize. "Thou shalt not make killable," she wrote in that prior book.[26] Speaking as a trained lab biologist, she saw with unique clarity how there is no rationalizing away or escaping the killing we perpetrate, the suffering we inflict (albeit not with equal complicity and not under conditions of our choosing). Rather than cultivate guilt, Haraway says we must stay with our responsibility for it. We must promote response-ability by "sharing suffering" every time, even if and when we decide to kill.[27] Because this articulation of the bloody fusion between politics and ethics has always struck me as extraordinarily fruitful and revelatory in an everyday as well as world-historic sense, I never allowed my worry that Haraway might prefer animals to humans to deter my deep gratitude (reverence, even) for her gifts. Until now.

Witness this diffident wish where Haraway is reflecting on the world "over a couple hundred years from now" and writes in a hopeful mode that "maybe the

25. *Donna Haraway: Storytelling for Earthly Survival* (2017), dir. Fabrizio Terranova, 90 min., earthlysurvival.org.
26. Donna J. Haraway, *When Species Meet* (Minneapolis: University of Minnesota Press, 2008), 80.
27. Haraway, *When Species Meet*, 69.

human people of this planet can again be numbered 2 or 3 billion or so."[28] In one faux-innocent speculative sentence, Haraway here disappears billions from her own eleven billion+ projection of this century's likely peak homo sapiens head-count.[29] Elusive as its explicit appearances turn out to be, in the final analysis, the numerical goal of population reduction undergirds and drives the entire book— and not just its pivotal chapter: Chapter 4—"Making Kin."[30] *Don't make babies* (as much as make kin) becomes the take-home injunction for the reader. The vision of trans-species *Gemeinschaft* that emerges is not so much post- as antihuman.

It is a vision that emerges shyly and guiltily rather than responsibly. In the Introduction, she calls "make kin not babies" a "plea" and dives right into a ten-dentious process of marshalling unwilling allies to her cause before she has even stated what it is. Indeed, what is most striking throughout is this guilt—Haraway's apparent discomfort with what she has to say, indeed, her near inability to say it— and this, in a way, is what says it all:

> For excellent reasons, the feminists I know have resisted the languages and policies of population control because they demonstrably often have the interests of biopolitical states more in view than the well-being of women and their people, old and young. Resulting scandals in population control practices are not hard to find. But in my experience, feminists . . . have not been willing to address the Great Acceleration of human numbers, fearing that to do so would be to slide once again into the muck of racism, classism, nationalism, modernism, and imperialism.[31]

The claim that any number of humans is expendable for the sake of the kin-community is advanced via a series of disavowals followed by oddly timid pieces of commonsense. I know what you're going to say . . . she repeatedly parries with

---

28. Haraway, *Staying with the Trouble: Making Kin in the Chthulucene* (Durham: Duke University Press, 2016), 103.

29. It is often the case that successful feminist struggle results in fewer human births and I have no problem saying that fewer human births will almost certainly be a planet-cooling (thus, "good") effect of the former goal. Common sense dictates that it would probably be easier to enjoy finite planetary resources in real communist abundance if there were three billion of us instead of seven or eleven billion (I dispute, however, that this is really fully knowable). It's still a far cry from that observation to a politics that takes population reduction as its end, even if it stringently avoids the language of "overcrowding" or even "overpopulation," as Haraway does.

30. Actually, the majority of Chapter 4 seems only to appear—as though spooked—in its own vast and apologetic footnotes. "For our people to revisit what has been owned by the Right and by development professionals as the "population explosion" can feel like going over to the dark side," she repeats in one of these, "[b]ut denial will not serve us." She knows, she says, that "reemphasizing the burden of growing human numbers, especially as a global demographic abstraction, can be so dangerous." And naturally, "coercion is wrong at every imaginable level in this matter, and it tends to backfire in any case, even if one can stomach coercive law or custom (I cannot)." Still, "[w]hat if nations that are worried about low birth rates (Denmark, Germany, Japan, Russia, Singapore, Taiwan, white America, more) acknowledged that fear of immigrants is a big problem and that racial purity projects and fantasies drive resurgent pronatalism?" [Haraway, *Staying with the Trouble*, 209.]

31. Haraway, *Staying With the Trouble*, 6.

"but . . ." For example: "But that fear is not good enough . . . a nine billion increase of human beings over 150 years, to a level of 11 billion by 2100 if we are lucky, is not just a number; and it cannot be explained away by blaming Capitalism or any other word starting with a capital letter."[32]

For Haraway now, it seems that what is bad is "scandals in population control practices," not population control per se, even if historically the two can hardly be called distinct. While that last catty sentence sticks in the craw, it is barely the worst of what's here. If 9,000,000,000 is indeed not "just a number" (it certainly seems that way to me), Haraway declines to tell us what else exactly it is. Leaving the implication open, she introduces—as though it were already legitimated—the word "urgency" as a synonym for projected population increased, without nailing down what the emergency consists of. Of course, what Haraway analyzes as "avoidance," based in "fear" (of sliding into the muck of "racism, classism, nationalism, modernism, and imperialism. . . .") could also be given the benefit of the doubt and interpreted as a decision, a conscious rejection.[33]

With "make kin, not babies," Haraway is far from the first to appreciate the seeming paradox and important truth: that making larger families might result in a smaller total population. That is, family enlargement can be a qualitative rather than quantitative matter. There is a class struggle already underway around the biological dimensions of the making of a good life—a struggle waged (among others) by abortion activists, single mothers, and commercial gestational surrogates threatening strike action. But rather than work through the preconditions and likely strategies for achieving (non-)reproductive justice politically, Haraway proceeds on the vague and simplistic presumption that the "kinnovations" of queer "oddkin" are necessarily better and less violent than biogenetic forms of family.

The blurring of descriptive and prescriptive elements is a poor replacement for dialectic immanence. Should a reader pause to ask, skeptically, what is politically "better" about "tentacularity," exactly?[34] They may not find a substantive answer. Not making babies is never much related to the objective of building counterpower. And if all of us "share flesh" already, what is the political purpose of fostering more flesh-sharing? Even if universal flourishing is easier to imagine when fewer

---

32. Haraway, *Staying With the Trouble*, 6–7.

33. There are, of course, numerous ways of avoiding structures of innocence that do not lapse into such all-too-easy dismissals. See, for example: Jasbir K. Puar, "'I Would Rather Be a Cyborg than a Goddess': Becoming-Intersectional in Assemblage Theory," *PhiloSOPHIA* 2, no. 1 (2012): 49–66; Jackie Wang, "Against Innocence," in *Carceral Capitalism* (South Pasadena, CA: Semiotext(e), 2018).

34. Donna Haraway, "Tentacular Thinking: Anthropocene, Capitalocene, Chthulucene," *e-flux journal* 75 (September 2016), https://www.e-flux.com/journal/75/67125/tentacular-thinking-anthropocene-capitalocene-chthulucene/.

humans are in the picture, desiring fewer humans is a terrible starting point for any politics that hopes to include, let alone center, those of us for whom making babies has often represented a real form of resistance.

This ethnocentric antimaternal impulse is an especially disappointing about-face for Haraway. "A Cyborg Manifesto" vindicated a non-innocent, anticolonial maternity (regeneration) symbolized by mutant or surrogate pregnancy. The cyborg, Haraway memorably declared back then, "is outside salvation history . . . it has no truck with . . . unalienated labor, or other seductions to organic wholeness . . ."[35] Compare this with the first pages of *Staying with the Trouble*, which replicate this prophetic tone, and even elements of its content: "Chthonic ones are beings of the earth, both ancient and up-to-the-minute. [They] have no truck with sky-gazing Homo . . . no truck with ideologues. . . . Chthonic ones are not safe. . . . They make and unmake; are made and unmade."[36] To those in the know, this is instantly recognizable music. But for others, it falls flat. To her readers hungry for mobilization and organization, as such, the bulk of *Staying with the Trouble* is likely to feel like a bit of a warming-over of previous themes and tendencies: cyborgicity wrung clean of systemic analysis and socialism, repackaged as a vague and omnipresent tentacularity.

While waxing forth about "symchthonic" potency, Haraway will usually mention the work of both making and unmaking, tying and cutting, and so on. But in practice she nowadays underemphasizes the potentially antagonistic-sounding acts of cutting and unmaking almost to the point of silence, even as she cuts humanity down to size. Her earlier call for new grapplings with the form labor takes as social domination (straddling species lines—*Biocapital*, Volume I) also seems forgotten. The core impetuses now appear to be downright *pro-work*, erasing her own gendered and others' companionate labor. What is left is a wholesome anti-laziness wedded to the injunction to be always doing: respond, act, cultivate, invent, discover, bind, work, be ever more capable and alert.

In the end, it seems that nobody at all except Haraway herself and Cayenne (her aging but "sporty" dog) is proactive enough. Even if her criticisms of noxious narratives, despondent or naive, hit their mark, it seems nowadays Haraway's biggest problem is having fallen out of love with the human masses. She suggests pessimistically that for future historians, "the period between about 2000 and 2050 on earth should be called the Great Dithering." This "Great Dithering," a periodization

35. Haraway, *Manifestly Haraway*, 8.
36. Haraway, *Staying with the Trouble*, 2.

drawn in part from the science-fiction author Kim Stanley Robinson, she proclaims will have been, "a time of ineffective and widespread anxiety about environmental destruction, unmistakable evidence of accelerating mass extinctions, violent climate change, social disintegration, widening wars, ongoing human population increases due to the large numbers of already-born youngsters (even though birth rates most places had fallen below replacement rate), and vast migrations of human and nonhuman refugees without refuges."[37] Just like that, she conveys her rather brutal certainty that humans, overall, are "dithering" *and will be for another thirty-three years.*

HORROR AND HOMONYM

Even those of us who have not read any H.P. Lovecraft are likely to have some familiarity with the death-cult god for which that prolific 1920s pulp science-fiction writer became famous. The weird, faux-arcane sound of the word "Cthulhu" has a widespread ability to conjure images of apocalypse and perhaps piles of skulls. A cursory scan of scholarship on Lovecraftian literature suggests a stable consensus that the Cthulhu Mythos was the vehicle of a genocidal fever-dream and obsessional racism. While serious fans and Lovecraft nerds still energetically debate the moral parameters of such creaturely invention, the media life of Cthulhu proceeds largely outside of their control. With this in mind, a wonderful review of *Staying with the Trouble* at the blog *Savage Minds* is authored by the Dread Destroyer (Cthulhu) himself:

> Sure my methods are 'controversial' but [Haraway] and I have the same goal in mind: confronting our shared ecological crisis by addressing the problem of accelerating human population growth. Whereas she seeks to carve out the possibility that feminism can navigate the racist and eugenicist histories of limiting human reproduction, I advocate for a strategy of direction action, i.e., human sacrifice.
>
> Haraway mistakenly believes she has inoculated herself against my minions by adding a superfluous 'h' to Cthulhu. . . . I am skeptical that she did not mean to summon me by speaking my name.[38]

Indeed.

---

37. Haraway, *Staying with the Trouble*, 144–45.

38. Matt Thompson, "Staying with the Trouble: Making Kin in the Chthulucene [Review]," *Savage Minds* (blog), November 18, 2016, https://savageminds.org/2016/11/18/staying-with-the-trouble-making-kin-in-the-chthulucene-review/.

Haraway protests that "Cthulhu . . . plays no role for me."[39] An unintentionally comic apophasis. And she instructs us repeatedly to "note the different spelling" as we approach her "Chthulucene": Cthulhu/Chthulhu. "Cthulhu (note spelling), luxuriating in the science fiction of H.P. Lovecraft, plays no role for me, although it/he did play a role for Gustave Hormiga, the scientist who named my spider demon familiar. . . . I take the liberty of rescuing my spider from Lovecraft for other stories."[40] In actuality, it is the spider (not the sublime misanthropic domination of Cthulhu) and the Indigenous cosmologies (not Lovecraft) who have been marginalized in this book. Haraway's forced insistence that something she has just named "plays no role" is an almost ridiculous moment from a self-proclaimed "material-semiotic" thinker. How many readers would spot the difference without being told to in the footnotes? How many would imagine it to be remotely significant? She ropes the Dread Destroyer (negatively) into her concept herself, so how can Cthulhu be quite so "irrelevant" to Chthulucene chthonic ones as all that? How, in other words, does one "rescue" a concept from a tradition—as many antiracist science-fiction authors have done for H.P. Lovecraft—without, in some sense, assessing that tradition?

Surprisingly, in the documentary film *Donna Haraway: Storytelling For Earthly Survival* (2017) by Fabrizio Terranova, Haraway tries to suggest that the Chthulucene is "a kind of joke" because "it too threatens to become too big" of a concept, just like the Anthropocene concept she is critiquing. In my view, though, it is a joke that misses badly; a lapse in judgment that is almost slightly shocking. By the same token, after being challenged on it in *The London Review of Books*, Haraway described the book's title framework has an "unconscious aural pun."[41] But a 2013 interview with Haraway by Martha Kenney features a fascinating (and definitely conscious, rather than unconscious) discussion of "the Lovecraft story": one that could generatively be put into conversation with afropessimist and decolonial scholarship on kinship, survival, and reproductive justice in the wake of actually existing genocides.[42] To paint with a homonym of H.P. Lovecraft's murdering titan represents a choice: one that, were it not disavowed, could be really interesting. My argument, to be clear, is by no means that Haraway should not have 'touched' Lovecraft. Indeed, far more than whether Lovecraft plays a role, the point

39. Haraway, *Staying with the Trouble*, 173.

40. Haraway, *Staying with the Trouble*, 173.

41. Donna Haraway, "Cyborgs for Earthly Survival!," *London Review of Books* 39, No. 13 (June 29, 2017), sec. Letters, https://www.lrb.co.uk/v39/n13/letters.

42. Donna Haraway and Martha Kenny, "Anthropocene, Capitalocene, Chthulhucene: Donna Haraway in Conversation with Martha Kenney," in *Art in the Anthropocene: Encounters Among Aesthetics, Politics, Environments and Epistemologies*, eds. Heather Davis and Etienne Turpin (London: Open Humanities Press, 2015), 255–69.

is, clearly, *what* role.[43] Haraway is the one to have taught us that semiosis, like biology, is never innocent; that—borrowing from Marilyn Strathern—"it matters which stories tell stories, which concepts think concepts."[44] Haraway herself, meanwhile, now "wishes" she had instead have opted for "Chthonocene."[45]

As one would expect of this other titan (of socialist-feminist antiracist thought), Haraway makes several gestures declaring her cognizance of systemic colonialism, capitalist austerity, white supremacy, and their manifestations in the form of reproductive stratification. She notes (correctly) that many people she holds dear "hear neo-imperialism, neoliberalism, misogyny, and racism in the 'Not Babies' part of 'Make Kin Not Babies'"—she even comments, "who can blame them?" in parentheses.[46] Nevertheless, in a thoroughgoing evasion of these issues, the reduction of human population imagined by Haraway takes place in the context of a racially unmarked (i.e., white) community situated at Gauley Mountain in West Virginia. Here, the parable tells, thanks to chains of events set in motion by compostists, "human numbers . . . were declining within a deliberate pattern of heightened environmental justice" by the year 2220. Gauley Mountain is the current real-life home of the white "ecosexual activists" who co-directed *Goodbye Gauley Mountain: An Ecosexual Love Story* (2013): the radical sex-educator Annie Sprinkle and the artist and filmmaker Elizabeth Stephens (who also directed *Water Makes Us Wet: An Ecosexual Adventure*, 2017). Sprinkle and Stephens are, presumably, the template "compostists" in question.[47] As detailed in one article about their project, "Ecosexuals in Appalachia," Sprinkle and Stephens returned to Stephens' hometown of Montgomery, West Virginia, in order to engage with the local LGBTQ community and examine Appalachians' responses to the ecocidal practice of mountain-top removal.[48]

43. To reiterate, our point is not that deploying 'Cthulhu' is in itself necessarily racist. On the contrary: while Cthulhu is a fictional figue who decimates humanity, it was not so much his presence in the text as the author's odd, oscillating avowal/disavowal of him ("Cthulhu plays no role for me") that caught our notice and prompted concern, given its tie to a call for population reduction from eleven billion to two billion. Our problem is with the relation between annihilationist science fiction and the so-called "burden of human numbers," the matter declared core by Haraway yet symptomatically hidden in her footnotes.

44. "It matters what ideas we use to think other ideas (with)." Marilyn Strathern, *Reproducing the Future* (Manchester, UK: Manchester University Press, 1992), 10. Referenced in Haraway, *Staying with the Trouble*, 12, 14, 34, 35, 37, 39, 57, 94, 96, 101, 118, 165.

45. Haraway, Letters, *London Review of Books*.

46. Haraway, *Staying with the Trouble*, 208.

47. As with "kinnovation," the language in which Haraway now couches her interventions would not seem out of place in Silicon Valley. Transhuman mutations and butterfly-beards notwithstanding, the question arises: is the "Communities of Compost" an image of genuinely transformative process or a kind of polyamorous start-up? [See Haraway, *Staying with the Trouble*, 138.]

48. Cynthia Belmont, "Ecosexuals in Appalachia: Identity, Community, and Counter-Discourse in Goodbye Gauley Mountain," *ISLE: Interdisciplinary Studies in Literature and Environment* 25, no. 4 (2018): 742–66.

I have no beef with the Appalachian ecosexualists. But they do not represent, in class or racial terms, the migrant, Indigenous and Black proletarian constituencies that a piece of speculative fiction about ecological liberation in America should, obviously, center. In Haraway's stated vision of the future, after all, the "pattern [of heightened environmental justice] emphasized a preference for the poor among humans, a preference for biodiverse naturalsocial ecosystems, and a preference for the most vulnerable among other critters and their habitats. The wealthiest and highest-consuming human populations reduced new births the most... but human births everywhere were deliberately below replacement rates."[49] Given its combination of white poverty and (on the part of the ecosexualists) relative white affluence, then, Gauley Mountain should have seemed like an odd choice for the Camille Stories' location. Undaunted, Haraway repeats that if I am appalled by her grasping the nettle of "the" population question (or, as she puts it, the issue of the Great Acceleration of human numbers), then I might be suffering from "beliefs and commitments... too deep to allow rethinking and refeeling," comparable to the "Christian climate-change deniers who avoid the urgency... because it touches too closely on the marrow of one's faith."[50] How deep, precisely, should our commitment to antiracism be?

There is no such orthodoxy, no such denial, when it comes to proletarian (especially Black and Brown people's) fertility rates. These have long been conceptualized as a threat and a problem, including within feminism. On the contrary, critical demographers still have to fight hard to bring gross structural inequalities—in mortality rates rather than fertility—into the frame at all. If Haraway were really "rescuing" and recuperating images of degeneracy (what James Kneale calls Lovecraft's core topoi of racial "contamination" and "infection"[51]) for the purposes of antiracism, wresting them away from fascist mythmaking, she would need to carefully center an analysis of the centrality of border-policing and population discourse to white supremacy. She does not do this. She expresses an appetite for a "wormy pile" of "chop[ped] and shred[ded] human as Homo," a banquet of "Humanity as humus," but without tackling the border regimes that fatally control and limit this supposedly joyful "diverse" commingling.[52] "Living-with and dying-with each other potently in the Chthulucene can be a fierce reply to the dictates of Anthropos and Capital," she blithely remarks.[53] If it can, it isn't really clear how, or for whom, this is true. In short, Haraway is trafficking irresponsibly in racist narratives.

49. Haraway, *Staying with the Trouble*, 159.
50. Haraway, *Staying with the Trouble*, 208.
51. James Kneale, "From Beyond: HP Lovecraft and the Place of Horror," *Cultural Geographies* 13, no. 1 (2006): 106–26.
52. Haraway, *Staying with the Trouble*, 32 and 160.
53. Haraway, *Staying with the Trouble*, 2.

The cyborg stood—quite shockingly—for a politics of "pollution."[54] Insofar, I remain for reading Haraway against Haraway. For all her chastisement of "bitter cynicism," and for all her talk of mud and piss and worms, the chanting goddess who has displaced the earlier cyborg, at least in the pages of *Staying with the Trouble*, is too much of a clean-living misanthrope—and above all, too much of a pessimist—to be a comrade.

Meanwhile, her neglected (if not disavowed) framework of cyborgicity becomes a more and more potent heuristic for thinking class composition and embodying its struggles every day. "Cyborgs for Earthly Survival!" was the slogan Haraway submitted to *Socialist Review*. That spirit still lives in the interstices of *Staying with the Trouble*. Part of our task is indeed "not to forget the stink in the air from the burning of the witches, not to forget the murders of human and nonhuman beings in the Great Catastrophes named the Plantationocene, Anthropocene, Capitalocene."[55] Part of it is, indeed, to "move through mourning to represencing," to grow capable of response, to become kin, and to "stay with" trouble. But the main thing is to make an altogether bigger kind of trouble.[56]

Tentacular, spidery aesthetics are all well and good, but they do not escalate struggle. These vague "chthonic" signifiers of well-meaning are a flimsy challenge to their namesake, the Great Old One, Cthulhu—that vivid necro-patriarchal savior-figure who is a caricature, arguably, of imperial capital. It is as though these new Harawayian companions—the chthonic ones—are making precisely the error she bewailed over the cyborg: "wallowing in the sublime of [His] domination."[57]

I have been relieved to see that, in the interviews in Terranova's aforementioned film, Haraway contradicts some of the elements of *Staying with the Trouble* laid out in this essay. "It is really important to be in revolt," she emphasizes there. "We do have to practice war. We do have to be for some worlds and against others."[58] Ultimately, this Cthulhu guy's got nothing on cyborg revolution when it comes to abolishing present realities. What if the cyborgs made a comeback? They knew who their enemies were. Overpopulation did not number among them. There is so much on Earth, after all, that we really do have to destroy.

54. Haraway, *Manifestly Haraway*, 57.
55. Haraway, *Staying with the Trouble*, 166.
56. Haraway, *Staying with the Trouble*, 166.
57. Donna Haraway and David Harvey, "Nature, Politics, and Possibilities: A Debate and Discussion with David Harvey and Donna Haraway," *Environment and Planning D: Society and Space*, vol. 13 (1995), 507–27, 514.
58. Donna Haraway, in *Donna Haraway: Storytelling for Earthly Survival* (dir. Terranova).

POSTCAPITALIST ECOLOGY

A COMMENT ON INVENTING THE FUTURE

First published November 2015

Published in 2015, Nick Srnicek and Alex Williams' (henceforth S&W) Inventing the Future: Postcapitalism and a World Without Work proved to be a controversial and generative text for the Left. The book offers a critique of existing left strategy and especially its romantic attachments to certain forms of struggle, such as symbolic marches, direct actions, unmediated localist politics. Fear of the modern, it would seem, has given way to fear of the future as such. Such an idea it could be argued actually prevents the working class from abolishing itself, putting it in a reactionary and defensive posture rather than a creative and expansive one. S&W propose instead a series of maneuvers to put the Left in front again: *most excitingly, perhaps, is an embrace of automation as a step to post-work politics. What are the implications and blind spots of such an argument, especially with regards to ecological crisis, unaddressed in the book? OOTW member Joseph Kay investigates in this review.*

Inventing the Future was also reviewed by OOTW members Sophie Lewis and David Bell. Their more critical review focuses on its inattention to colonial relations of extraction, its refusal of Indigenous and decolonial struggles and knowledges, and the manner in which its critique of 'moralizing' leftisms fails to account for the need for accountability in struggle.[1]

---

1. Sophie Lewis and David M. Bell, "(Why We Can't) Let the Machines Do It: A Response to Inventing the Future," *The Disorder of Things*, https://thedisorderofthings.com/2015/11/05/why-we-cant-let-the-machines-do-it-a-response-to-inventing-the-future.

The following comment on *Inventing the Future* may read more critically than I expected. In mitigation, I should say that I'm onboard with many of the key themes of the book. I am wholly sympathetic to antiwork politics, generally in favor of automating away toil (with qualifications which will become apparent) and agree that the replacement of global capitalism requires scalability, comfort with complexity, long-term strategy, utopian imagination, and a plurality of organizational forms and infrastructures.

The critical tenor of what follows arises less from disagreement than from my attention to what appear to be the ecological silences in the text. In particular, I focus on the implied conception of nature imported through S&W's adoption of an avowedly modern rhetoric of progress and control. I continue to investigate some of the unmentioned ecological premises of both the project of full automation and their more general contention that "we are usually not better off taking the precautionary path."[1] My argument is not to reject a high-tech, low-work future, but to outline some of the problems to be addressed in rendering such a "hyperstitional" image ecologically livable.

MODERNITY AND THE IDEOLOGY OF NATURE

Early on in *Inventing the Future*, S&W summarize their thesis: "If complexity presently outstrips humanity's capacities to think and control, there are two options: one is to reduce complexity down to a human scale; the other is to expand humanity's capacities. We endorse the latter position."[2]

Read in an ecological light, the conjunction of "think and control" affords two readings. The first and more obvious situates their argument within what Neil Smith called "the ideology of nature." Smith argued the ideology of nature had two poles. The first, a modernizing politico-theological argument for scientific progress as the means to conquer and subdue nature. Here, the imaginary is mechanical, and separation from—as dominion over—nature is understood as an emancipatory process.

The second pole was romanticism, which sought rather to revere nature and tread lightly on its arcadian wilderness. This emerged as a countermovement to modernism. As Smith puts it, "the romanticization of nature was not even possible until nature had already been substantially subdued. . . . One does not pet a

---

1. Nick Srnicek and Alex Williams, *Inventing the Future: Postcapitalism and a World Without Work* (London and New York: Verso Books, 2015), 177.
2. Srnicek and Williams, *Inventing the Future*, 16.

rattlesnake until it has been de-fanged."[3] Here the imaginary is organic, and separation from nature is understood as the loss of an originary wholeness. S&W argue that contemporary left politics is mired in such a tendency toward romanticization of prefigurative events, novel situations, and "the small-scale, the authentic, the traditional, and the natural."[4] They dub such a tendency "folk politics."

In many ways, S&W's human-scale folk politics appear aligned with Smith's back-to-nature romanticism, while S&W's own position appears to be a reassertion of modernist ideology against romantic backsliding. Their vocabulary of progress and modernity certainly draws its rhetorical force from this tradition. When they lambast folk politics for valorizing "feeling over thinking" or insist that "as we acquire . . . scientific knowledge of the natural world. . . we gain greater powers to act," and declare that there is "no organic wholeness to be achieved. Alienation is a mode of enablement," they appear firmly within this camp.[5]

Yet through the lens of Smith, both poles of the ideology of nature have problematic premises and owe more to one another than either would like to admit: "hostile or friendly, nature was external; it was a world to be conquered or a place to go back to."[6] Both are premised on the contradictory dualism of an external nature to be conquered or revered, and a universal nature including "human nature" (with its "savage" component to be civilized or reconnected with). Smith links this contradictory dualism to the historical development of capitalism, through colonialism and industrialization, where nature (external and human) really appears as a frontier to be conquered and an input into the production process.[7] This production process in turn reproduces the conditions of the ideology of nature. Hence the contradictory ideology of nature is the inverted reflection of *capitalist* modernity.

In Chapter 4 on "Left Modernity," however, S&W are keen to differentiate their project from capitalist modernity and its colonialist, teleological conceptualizations of progress. Hence a second reading is also possible. Environmentalists often mistake instrumental goods (the local, hard work, organic food) for intrinsic goods, and reify a pristine nature (anarcho-primitivism/anticivilization being the *reductio ad absurdum* of this tendency). Modernists just as frequently fetishize technological fixes, disavow practical knowledges, and champion a hubristic image

3. Smith, *Uneven Development*, 26.
4. Srnicek and Williams, *Inventing the Future*, 10.
5. Srnicek and Williams, *Inventing the Future*, 11, 81, 82.
6. Smith, *Uneven Development*, 21.
7. See also Moore, *Capitalism in the Web of Life*.

of modernizing scientific progress at odds with the caveats and qualifications that permeate a typical scientific paper (one need only think of "The Ecomodernist Manifesto" here).[8]

Where both modernist and romantic (folk) positions take nature as a given, Smith's critique of the ideology of nature takes aim at its premises: capitalist social relations and the inverted image of nature they reproduce. In its place is the notion of the *production of nature*, which crucially "implies a historical future that is still to be determined."[9] Here we seem close to what S&W want to argue, but to make this reading work their conflation of knowledge and control needs to be teased apart. Put simply, knowledge *does not* imply control, and sometimes—in certain classes of complex systems—knowledge provides a negative proof of such a possibility.

Climate change is a case in point wherein extensive knowledge of nature has not implied control. The greenhouse gas emissions driving global warming are uncontroversially *produced*. Yet their consequences are neither intended nor wholly predictable, let alone controlled. The production of nature includes everything from the most megalomaniacal global geoengineering scheme to the most low-tech localist permaculture. In shifting away from the dualist framing (think: ecological footprints, as if humanity impressed itself from outside the planet), it moves away from the either/or framings that characterize much environmental debate—and indeed S&W's approach quoted at the start of this section. Instead, more productive questions of *how* we produce nature (including ourselves) are opened without presuming an answer in advance.

This seems like a necessary move if we are, as S&W urge, to treat the universal not as a given content but "an empty placeholder that is impossible to fill definitively."[10] However, it may pull the argument in directions too folkish for S&W's tastes. One gets the impression they would be enthusiastic to learn of a self-replicating, carbon-scrubbing machine with the potential to geoengineer the climate but disappointed to learn this machine is called "forest." (Conversely, folk partisans would bristle at hearing life described in such machinic terms!) When S&W embrace "emancipatory alienation," they do so to reject the existence of any originary wholeness to which it is possible to return. But this remains a dualist frame which suggests that because no such pristine moment exists, nature must

8. Asafu-Adjaye et al., "An Ecomodernist Manifesto."
9. Smith, *Uneven Development*, 48.
10. Srnicek and Williams, *Inventing the Future*, 77.

be separated from and controlled. This rhetorical move forecloses many more generative possibilities.

S&W mention in passing Donna Haraway's "Cyborg Manifesto," which one might think could fall on the "modernist" side of things. Haraway's cyborgs "are wary of holism, but needy for connection."[11] This is not an embrace of emancipatory alienation or simple technological augmentation of the human, but an explicitly anti-dualist ontological claim that there is *neither* an originary wholeness *nor* clear boundaries (and hence separation/alienation) between the human, the natural, and the technological. Indeed, Haraway identifies both original wholeness and alienation as two moments of the same dualistic "Western epistemological imperatives" she is seeking to subvert.[12] The figure of the cyborg calls into question the binary distinction between the mechanical and the organic.

While the romantic (and folk?) critique of modernity is that it is a loss of a primordial Arcadia, a separation to be undone, there are further critiques of modernity that do not fall into these same trappings. Particularly evident are postcolonial critiques of modernization theory, which argue that "modernity" tends to be defined through the disavowal and rejection of the "traditional."[13] In *Inventing the Future*, folk politics often seems to play this role of disavowed other, against which progress, modernity, and the future can be defined. Yet in the introduction, S&W also insist that "folk politics is necessary but insufficient."[14] Taking this caveat seriously points more to a project of connection than alienation, where "on-the-ground knowledge must be linked up with more abstract knowledge."[15]

This connective theme evokes James C. Scott's famous—and in S&W's terms, folk—critique of "authoritarian high modernism," whose failures he diagnosed as arising from privileging simplified abstractions to the exclusion of practical local knowledge.[16] In an approving comment on the seventies Chilean attempt at computer-assisted economic planning known as Project Cybersyn, S&W describe it as a form of *bricolage*: improvising something new from the materials at hand.[17] Yet the *bricoleur* does not control and conquer nature. As Scott and others have argued, it is precisely such supple, inventive relations which suffer under modernization

11. Haraway, *Manifestly Haraway*, 9.
12. Haraway, *Manifestly Haraway*, 58.
13. Gurminder K. Bhambra, *Connected Sociologies* (London: Bloomsbury Publishing, 2014); Walter Mignolo, *The Darker Side of Western Modernity: Global Futures, Decolonial Options* (Durham and London: Duke University Press, 2011).
14. Srnicek and Williams, *Inventing the Future*, 12.
15. Srnicek and Williams, *Inventing the Future*, 174.
16. Scott, *Seeing Like a State*.
17. Srnicek and Williams, *Inventing the Future*, 149.

schemes. Instead, the bricoleur occupies a more cooperative, connective, and pragmatic relation to their surroundings. In the words of Indigenous knowledges scholar Daniel Wildcat, "Today, the problem is that the measure of technological progress is often thought of as the extent to which humankind can control and mitigate the so-called forces of nature. I find it hard to imagine a more problematic and potentially dangerous idea. We must figure out a way to live with nature."[18]

Wildcat gives the example of disastrous attempts by the US Army Corps of Engineers to control the Missouri and Kansas rivers to protect cities built on flood plains, which resulted in two five-hundred-year floods and hundreds of millions of dollars in damage in a three-year period in the nineties. While warning against the rejection of modern science and technology, Wildcat insists we "face the challenge of identifying technologies that have value beyond the exploitative narrow economic measures of profit."[19] Against the mentality of control, he advocates a cooperative approach that makes use of place-based knowledge to construct cities and infrastructure out of danger in the first place, rather than relying on expensive and ultimately counterproductive engineering projects. Little about this system resembles the mode of folk politics described by Srnicek and Williams.

It may be that S&W are simultaneously making the general argument about the nature of the universal (an empty placeholder whose content we must contest), and a specific argument for their favored content (a project of emancipatory alienation and control of complexity). But if so, failing to make this clear means the latter position tends to foreclose the opening intended by the former. Skepticism in advance for positions such as Wildcat's "Indigenous realism" contributing to the collective project of inventing the future would seem like a mistake. Such a result brings into focus modernity's tangled relationship to colonialism.

MODERNITY AND COLONIALISM

S&W are at pains to distinguish their advocacy of progress, modernization, and the future from the colonial history this rhetorical palette evokes. Indeed, having recounted their caveats against teleology, unilinearity, and eurocentrism (Chapter 4), it seems like this choice of rhetorical frame has as much to do with announcing a break with continental philosophy's received postmodern wisdom as a wholesale embrace of the European Enlightenment. After all, the latter engaged in scientific

18. Daniel R. Wildcat, *Red Alert!: Saving the Planet with Indigenous Knowledge* (Colorado: Fulcrum Publishing, 2010), 127.
19. Daniel R. Wildcat, *Red Alert!*, 128.

racism and "high-risk adventures" of colonial appropriation alongside its more defensible achievements.[20] Such a reading is supported by their call in the conclusion to "reappraise which parts of the post-Enlightenment matrix can be saved and which must be discarded," an approach more bricoleur than *revanchiste*.[21]

However, the relationship between modernity and colonialism is not a purely historical question but also a very material one. The automated technologies required for a post-work future are thoroughly premised on extractive relations of dominance which produce immiseration for "surplus populations" the world over, but especially those rendered so by colonialism. This is not to raise a primitivist objection that machines are inherently at odds with nature (such a dualistic frame having been already rejected), but to stress that the colonial imbrication with modernity cannot simply be disavowed. It has to be undone.

To take one example, rare-earth minerals are essential components of modern electronics and hence any automation project. Yet mining these minerals produces radioactive slurry tailings and refining them produces toxic acid byproducts. The environmental justice movement has long highlighted the environmental racism whereby exposure to toxic waste is unevenly distributed along lines of race and class.[22] Writing of Silicon Valley as high-tech industry's ground zero, Nick Dyer-Witheford observes that "on the one hand, palatial billionaire mansions, and on the other, 23 'Super-Fund' abandoned toxic waste sites scheduled for special clean up operations, the most of any county in the US."[23]

Within the global division of labor this takes a distinctly neocolonial form. The Democratic Republic of the Congo (DRC), for example, is one of the world's main producers of cobalt and coltan, continuing the violent history of extractive processes that runs from Leopold's genocide through the assassination of Patrice Lumumba to the present day. Coerced and slave labor has long existed in the DRC's mines. At the other end of the tech lifecycle is Agbogbloshie, a commercial district in the Ghanaian capital Accra, it hosts vast slums and an e-waste dump where pollutants including lead, mercury, arsenic, and dioxins are present in high concentrations. Thousands of people live amongst toxic waste.

20. Beate Jahn offers a persuasive account of the role of the colonial encounter in generating central categories of specifically modern political thought in *The Cultural Construction of International Relations: The Invention of the State of Nature* (London: Palgrave MacMillan, 2016).

21. Srnicek and Williams, *Inventing the Future*, 181.

22. Dorceta E. Taylor, *Toxic Communities: Environmental Racism, Industrial Pollution, and Residential Mobility* (New York: NYU Press, 2014).

23. Nick Dyer-Witheford, *Cyber-Proletariat: Global Labour in the Digital Vortex* (London: Pluto Press, 2015). See also David N. Pellow and Lisa Sun-Hee Park, *The Silicon Valley of Dreams: Environmental Injustice, Immigrant Workers, and the High-Tech Global Economy* (New York: NYU Press, 2002).

While automation could in principle minimize human exposure to toxic sub-stances within the labor process, it is currently premised on the disposability of racialized populations treated as less-than-human. The problem of toxic waste is not mentioned in *Inventing the Future*, yet without addressing head-on how this dependence could be undone, any project of full automation will be complicit in and dependent upon the continuation of colonial, racialized social relations.

Another example of the imbrication of modernity and coloniality is apparent in one of the proposed technological fixes to climate change: solar radiation man-agement (SRM). This is not a technology advocated in the pages of *Inventing the Future*, however it serves as a good illustration of what is at stake.

The principle behind SRM is simple. The warming aspect of climate change is proximally caused by outgoing longwave (infrared) radiation becoming trapped in the Earth's atmosphere, raising the surface temperature. To counter this, it is proposed to use atmospheric aerosols (or orbital reflectors) to intercept a portion of the Sun's incoming shortwave radiation (insolation). This would reduce the total energy input to the Earth system and offset the surface warming caused by green-house gases. SRM is thus exemplary of technologically augmented human control over complex systems.

The Earth's atmosphere is a prototypical chaotic system. This is to say it is deterministic but with sensitive dependence on initial conditions. Consequently, the localized consequences of SRM are not easily modeled, as SRM alters the atmo-sphere's temperature gradient, known as the lapse rate, which is critical to meteo-rology. Some models suggest that the main adverse consequence of SRM would be severe droughts in regions dependent on seasonal rains, principally Sub-Saharan Africa and parts of Asia.[24] S&W's critique of the precautionary principle could eas-ily be invoked by SRM advocates here:

> the precautionary principle contains an almost inherent lacuna: it ignores the risks of its own application. In seeking to err always on the side of caution, and hence of eliminating risk, it contains a blindness to the dangers of inaction and omission. While risks need to be reasonably hedged, a fuller appreciation of the travails of contingency implies that we are usually not better off taking the precautionary path. The precautionary principle is designed to close off the future and eliminate contingency, when in fact the contingency of high-risk adventures is precisely what leads to a more open future.[25]

24. See Naomi Klein, *This Changes Everything*, 256–90 for a critique of SRM. See David Spratt and Philip Sutton, *Climate Code Red: The Case for Emergency Action* (Carlton North, Vic: Scribe Publications, 2008), 257–66 for a measured advocacy of temporary SRM as an emergency measure.
25. Srnicek and Williams, *Inventing the Future*, 177.

What the SRM example highlights is that the human and nonhuman populations taking the risks and shouldering the consequences need not necessarily coincide. It's easy to wax lyrical about high-risk adventures when someone else picks up the tab. This "someone else" is determined through the extant relations of power, hence typically racialized and usually (post)colonial. It is clear from S&W's caveats that they are keen to dissociate their left modernity from such implications, but they also provide arguments which, in the absence of delimitation on their part, could readily be appropriated to such ends.

CONCLUSION

Srnicek and Williams adopt wholesale the rhetoric of modernization but attempt to distance themselves from its darker side. This is accomplished through both a series of caveats and qualifications, and by attempting to redefine the content of modernity. The conjunction of knowledge and control in their summary of their approach affords two readings. The obvious one operates within the ideology of nature and allies itself to the historical project of human separation from and control of nature, while disavowing the local and particular as a conservative, precautionary brake on progress, longing for a mythic originary wholeness.

However, their "necessary but not sufficient" framing of "folk politics" affords another reading that eludes this binary framework. Read through Smith's notion of the production of nature and Haraway's figure of the cyborg, S&W can be read as advocating a tactic of bricolage that brings together various practical and abstract knowledges in a cooperative production of nature. This reading might go against S&W's advocacy of emancipatory alienation, but it is perhaps more helpful for contemporary political movements.

While keen to distance themselves from modernity's colonial origins, the ongoing imbrication of the project of full automation with racialized and neocolonial relations produced in and through high-tech production must be addressed head on in any project attempting such rehabilitation. The default position in *Inventing the Future* is therefore one of complicity. One possible way to address this would be to generalize the deployment of industrial ecology/closed-loop production methods, whereby waste outputs are engineered to become inputs to other processes.[26]

26. These techniques already exist within capitalism, but deployment is constrained to those cases where minimizing waste helps maximize profits. See the film *Waste = Food* for examples of extant closed-loop production, available at vimeo.com/3237777.

This approach would constitute a bricolage of applied scientific and place-based knowledges aiming at a cooperative connection with, rather than control of, the material-energetic flows of wider ecological webs. It would espouse a production of nature that is neither a frontier to conquer nor an idyll to which one could return. We could even speculate that necessarily local anti-extractive struggles—relatively immune to capital flight—may catalyze the global development of closed-loop methods in much the same way that S&W hope that wage struggles will catalyze automation.

However it is addressed, the racialized, colonial premises of full automation cannot simply be disavowed, they have to be undone if the emancipatory potential of automation is to be universal. This argument could be seen as an extension of the book's repeated emphasis on the quietly constraining and enabling role of infrastructure.[27] This is particularly important given the potential for antiwork politics to bridge the red-green divide that too frequently allows jobs-and-growth trade unionists and environmentalists to be divided and ruled. S&W contend that "reductions in the working week would lead to significant reductions in energy consumption and our overall carbon footprint. Increased free time would also mean a reduction in all the convenience goods bought to fit into our hectic work schedules. More broadly, using productivity improvements for less work, rather than more output, would mean that energy efficiency improvements would go towards reducing environmental impacts. A reduction in working hours is therefore an essential plank in any response to climate change."[28] Yet such claims are frequently contradicted by the available evidence.

In the context of climate change, S&W make several arguments which seem too easily appropriated to support technofixes rather than the needed social-ecological transformation. Ecology is frequently invoked as metaphor—as in organizational ecology—but an ecological perspective doesn't appear as more than a fringe benefit to the program of full automation. Climate change demands a utopian politics against default dystopian despair. Inventing an antiwork ecological politics is surely necessary and desirable, and indeed, "doing so requires us to salvage the legacy of modernity and reappraise which parts of the post-Enlightenment matrix can be saved and which must be discarded."[29] I contend that what is to be discarded includes the ideology of nature, and modernity's blind spot to its own ongoing racialized/colonial imbrications.

27. Srnicek and Williams, *Inventing the Future*, 133.
28. Srnicek and Williams, *Inventing the Future*, 116.
29. Srnicek and Williams, 181.

IV

STRATEGIES

INTRODUCTION

ORGANIZING AMIDST CRISIS

The current situation of climate and environmental politics is largely chaotic, and yet has never been better for an organized communist response. On the international stage, the Paris Agreement has followed its predecessors in being largely ineffective. The world's largest states have elected reactionary leaders, many of whom appear delighted to watch the planet boil while imprisoning or disappearing their opposition. Every month seems to bring punctuated disasters and new evidence of accelerating trends.

This chaos has imbued the Green Left with new energy, some tendencies of which are quite hopeful. The blockade of the Dakota Access Pipeline on the land of the Standing Rock Sioux Nation, while ultimately unsuccessful in its main goal, highlighted and reinforced the power of a transnational movement for Indigenous sovereignty and land restitution. This movement has stalled the Keystone XL and Trans Mountain Pipelines for years: the latter was saved from the brink of failure after it was, astonishingly, purchased by the Canadian government to transport oil from the Alberta Tar Sands in 2016. Although previously quite marginal, an ecosocialist (and somewhat Keynesian) Green Left in the United States and United Kingdom has vigorously emerged in support of a "Green Marshall Plan" or "Green New Deal." Large-scale civil disobedience aimed at combatting ecological crisis has emerged around the planet, though some analyses, strategies, and slogans remain deeply insufficient.[1]

Within the Green Left, strategies for struggles against the state and capital are emerging from movements fighting border imperialism, global structural racial inequalities, neocoloniality and settler coloniality, and social reproduction struggles more generally. It is too early to say whether such movements will amount to a serious challenge for the state and capital. Out of the Woods has attempted to foreground what it might look like if ecological movements are not divorced from but are rather at the center of struggles against exploitation and domination, not as secondary effects of environmental crisis, but as primary elements of ecological crisis.

While already existing environmental justice movements around the world provide countless acts of inspiration, for various reasons these have yet to coalesce into a political project with the spatial, temporal or intersectional scope of the planetary ecological crisis.[2] There are reasons why many environmental justice movements have remained only ambivalently related to communist and anarchist projects. For example, the framing of the scales of action of some environmental justice movements romanticize the local, which requires a style of grassroots organizing that makes broadscale action difficult. The linking of elements of environmental justice struggle with radical Indigenous anticolonial and Black Liberation movements has made them targets of state and police action. Finally, the liberal frame of "justice" is all too easily incorporated into the state's reductive understanding of the demands of environmental justice as merely about distribution, participation, and recognition.

As demonstrated in Section II of this book, for Paul Kingsnorth, it is precisely the so-called failure of the alter-globalization movement that should return us to cherishing our local-national ecosystems. While the existence and growth of green nationalism and ecofascism must be ruthlessly critiqued and materially opposed, this cannot excuse the strategies of liberals and the Green Left as objects of critical analysis. Like Kingsnorth, the Green Left has at times been suspicious of coordinated international action, favoring instead softly localist frameworks like the Transition Towns movement or emphasis on small businesses, local food, and

1. Out of the Woods, "Extinction Rebellion: Not the Struggle We Need, Pt. 1," Libcom.org (blog), July 19, 2019, http://libcom.org/blog/extinction-rebellion-not-struggle-we-need-pt-1-19072019; Out of the Woods, "Extinction Rebellion: Not the Struggle We Need, Pt. 2," Libcom.org (blog), October 31, 2019, http://libcom.org/blog/xr-pt-2-31102019.

2. Laura Pulido, "Geographies of Race and Ethnicity II: Environmental Racism, Racial Capitalism and State-Sanctioned Violence," Progress in Human Geography 41, no. 4 (2017): 524–33.

community-scale energy development. All of these enclose a certain supposedly "natural" vision and scale of the social.

Yet communism in one country is as fictitious as green capitalism. If anything, the global environmental crisis demonstrates the supreme intractability of a local or national response alone. How then can struggles against capital, raciality, and coloniality resonate and coordinate in catastrophic times?

We wholeheartedly propose that the name *communism* could and should serve as a common point of reference for thinking about these struggles and their resonance across scales and histories. Haraway imagined "a political form that actually manages to hold together witches, engineers, elders, perverts, Christians, mothers, and Leninists long enough to disarm the state."[3] Our cyborg ecological approach applies to politics as much as natures. We continue to learn from migrant, queer, and prison-abolitionist political movements as well as from the Green Left and historical environmental justice movements. For example, thinking through what the Black Panthers called "survival pending revolution" helps us conceptualize how this coalition could work adjacent to state processes in ways that still build power towards an eventual overcoming of the state.

This is not an innocent position to take. Our politics and concepts emerge from disparate collective situations in the increasingly hollowed-out ruins of the Global North. But we also recognize that this might give us a situated and partial vantage point that could be helpful to comrades around the world.

This section begins with an examination of Naomi Klein's *This Changes Everything: Capitalism vs. the Climate*, a book published in 2014 that has been important in generating a more forcefully anticapitalist climate justice movement amid failures in international climate negotiations. In "Blockadia and Capitalism: Naomi Klein vs. Naomi Klein," we argue that much like climate justice movements worldwide, this book is riven with a contradiction concerning the relationship between the state and capital. If a global anti-fossil fuel movement—Blockadia—is to exist, why stop at demanding Keynesian band-aids? For our part, we want it all.

Who will be the subjects to bear such a movement? They do not yet exist at the scale and level of solidarity needed. In "Climate Populism and the People's Climate March," originally written in 2014, we investigate whether the People's Climate Movement might provide one possible organizing platform. Since that time, "left populisms" have become even more trendy, at least among certain sections of the commentariat. While we express some ambivalence about the possibilities of a left

3. Haraway, *Manifestly Haraway*, 16.

populism, we also think that, in their radical openness, such movements could easily be infiltrated, augmented, and turned about. It is not the language of "the people" *as such* that makes a movement, but instead the inventiveness towards expression of non-state and non-national forms of collective identification that do not quash difference. This, is solidarity.

As we argue in "*Après moi le déluge!*," the moment of the abolition of slavery in the 1860s US South was produced by a broad and contradictory field of political positions working within, adjacent to, and against capital. Together these provided excellent conditions for producing rapid expropriation and transformation of the capital relation. This essay, now five years old, would have benefited from deeper historical analysis as well as studies of and with contemporary prison-abolition movements and abolitionism beyond the United States, though we stand by its general point that drastic transformation is possible through a confluence of contradictory forces.

If communism is to exist, it will be created in a world riven by ecological crises, themselves wrought by racial capitalism. *Any* communism worthy of the name will now be disaster communism. By this, we mean that the revolutionary process of developing our collective capacity to endure and flourish will emerge within, against, and beyond ongoing capitalist disaster. Such communism is thus "the real movement which abolishes the present state of things" *and* the orientation towards a totality based not on the principle "from each according to [their] ability" but "to each according to [their] need."[4]

It must be emphasized: disaster communism will not be produced spontaneously *ex nihilo* by these terrible catastrophes. We do not wish catastrophe upon the world as a means to our end, but we must be realistic about the disastrous situation we find ourselves in. The key lesson we have learned from mutual-aid collectives that arise in response to disaster is the necessity of preparation, so that when the conditions are ripe, we are ready to move.

Why do we choose this word, communism? Why not stick with Solnit's "disaster communities," or the more palatable "disaster socialism" (as some within the Democratic Socialists of America have proposed) or perhaps "disaster anarchism" (as some communitarians suggest)? We do not always agree on this point.[5]

---

4. Karl Marx and Friedrich Engels, *The German Ideology: Part One, with Selections from Parts Two and Three*, ed. C. J. Arthur (New York: International Publishers, 1970), 57. Karl Marx, "Critique of the Gotha Program," in *The Political Writings* (London and New York: Verso Books, 2019), 1031.

5. See, for example, Alex's claim in our base magazine interview that 'I'd probably also go as far as to say that we should try to develop something else because I'm not even sure "disaster" is quite the right kind of word for encapsulating what we are really trying to resist and survive given that it's not one disaster or even a series

Communism will recall different movements and struggles in different places and times, not all of them savory.

For us, however, communism has a practical utility in naming our principled commitment to a transformative horizon and a refutation of the economic and political structure of the world as it currently is. Communism is not only a name for the struggle to negate the immiseration produced by capital, but is itself the name-under-struggle. To be under struggle is to be common.[6] Our essay "Disaster Communism: The Uses of Disaster" is our most recent elaboration of the utility of this name in the contemporary moment. These "disaster communities" are little glimpses of a world where social reproduction is orchestrated not through waged labor, commodities or private property, but through care, solidarity, and the passion for liberation and equality.

"Communism." Commitment to such a name might stave off recuperation by the state as well as differentiate us from the (national) social democrats. The twentieth century provided no shortage of evidence otherwise. Communities are nice, but in themselves have no politics; they can equally lend themselves to the reactionary *oikonomia* of the nation. And while we agree with anarchists on the question of state abolition, we worry both at the contemporary tendency to emphasize the individual, their moralist position, and an underestimation of the extra-state forces of capital.

To talk about strategy is not to present a teleological program. Instead, it is an oportunity to think through the momentary situations in which we see doors open and close. To ignore the specificity of these situations is to cosplay 1917, 1871, 1848, or 1791 over and over again. This is not to suggest history doesn't matter; as with climate change, the lessons of history do not exhaust the future. The ongoing and upcoming cycle of struggles can realize the promise of past revolutions—if we make it so. And as we have said all along, we want it all.

---

of disasters, it's a particularly potent mix of catastrophe and normality in which both are murderous' (page 37, above). After considerable discussion within our group we decided not to take this path, but rather to expand our definition of 'disaster' so as to account for this 'normality' as a disastrous condition.

6. Bini Adamczak, *Communism for Kids*, trans. Jacob Blumenfeld and Sophie Lewis (Cambridge, MA: MIT Press, 2017), 78. See also Jodi Dean, *The Communist Horizon* (London and New York: Verso Books, 2012).

BLOCKADIA AND CAPITALISM

NAOMI KLEIN VS. NAOMI KLEIN

First published January 2015

On its publication in 2014, Naomi Klein's This Changes Everything: Capitalism vs. the Climate *promised a renewed anticapitalist point of unity for a democratic movement against ecological crisis. This essay critiques the compelling and strange double-call of the book. Klein's dominant, major voice is reformist. Yet the book derives its drive from a second, minor call—a revolutionary invitation that underpins and threatens to undo the stable compromises of the first. To deliver reform, Klein depends upon a "people's shock from below." In response, we wonder why movements capable of generating such a shock should be content to settle for reforms. This essay turns Klein's minor call against her major voice, and accepts the book's invitation to change everything.*

On reflection, the critique we enact in this essay mirrors the tactics enacted in our politics. Like bricolage, we take what we need from Klein—a critique of the state, a desire for direct action, a suggestion of regeneration—and put it to uses perhaps unintended by the author. In the act of finding and reassembling these components, we give Klein more credit for a politics she might never avow or assert. We would argue it doesn't matter. The invitation remains; the call to think of a world beyond this one, built of fragments assembled from that which is already here.

TWO KLEINS

In her Twitter bio, Naomi Klein says she is "polarizing." In fact, the responses to her ecological magnum opus, the internationally bestselling *This Changes Everything: Capitalism vs. the Climate*, were almost universally enthusiastic. Though Klein's introduction frames it as "the hardest book I've ever written . . . because the research has led me to search out such radical responses," at the time of its release it was well-reviewed even by many who rejected the subtitle's signpost of radicality.[1] In *The New York Times*, Rob Nixon roundly praised the work, raising just one quibble: "what's with the subtitle? . . . Klein is smart and pragmatic enough to shun the never-never land of capitalism's global overthrow."[2] In his four-star *Telegraph* review, the former Conservative Member of European Parliament and World Bank employee Stanley Johnson (father of UK PM Boris Johnson) confidently stated that Klein is "no advocate of socialism." In the face of this unanimity, we seek not so much to return some polarity to Klein but to tease out polar contradictions present in *This Changes Everything*.

Klein's desire for a mass, popular climate justice movement pulls her in two directions. She calls for the creation of millions of green jobs, yet she advocates for liberating people from work. She advocates rapid fossil fuel abolition while arguing for a welfare state funded by taxes on fossil fuel profits. She takes aim at the profit motive yet endorses small local businesses as the fabric of the community. Rather than make accusations of confusion or hypocrisy, let's take seriously her claim to have been pushed into radically 'polarizing' positions by the urgency and severity of the climate crisis, and propose that two divergent Naomi Kleins—a "major Klein" and a "minor Klein"—have been formed who together make up the author of *This Changes Everything*.

The dominant voice is the major Klein, an idealist talking in terms of moral values who argues for an embedded liberalism and a redistributive state. Major Klein wants to create "plentiful, dignified work," not least of all for "families." She supports small local businesses, thinks of the children, and seeks to unite left and right to clean up corruption and get corporate money out of politics.[3] Major Klein tends towards explanations of social phenomena in terms of moral failings. The reckless pursuit of profit is a result of "greed," and so the economic crisis was

---

1. Naomi Klein, *This Changes Everything: Capitalism vs. The Climate* (New York: Simon & Schuster, 2014), 26.
2. Rob Nixon, "Naomi Klein's *This Changes Everything*," *The New York Times*, November 6, 2014, sec. Books, https://www.nytimes.com/2014/11/09/books/review/naomi-klein-this-changes-everything-review.html.
3. Klein, *This Changes Everything*, 125 and 118.

"created by rampant greed and corruption."[4] Its solution is a left-populist mobilization to support a Green New Deal.[5]

Against this we find minor Klein, her voice inchoately distributed across *This Changes Everything*. This Klein is a radical realist talking in terms of raw power and material interests. She stresses that "only mass social movements can save us now" and advocates for a "basic income that discourages shitty work."[6] Minor Klein proposes workplace occupations, blockades, and neighborhood assemblies. She hints at a regenerative politics of the nonfertile. And, she is disillusioned with even ostensibly enlightened leftist politicians, whether in the United States, Bolivia or Ecuador. While these two Kleins might be a readerly construct on our part, it is a sympathetic one that helps identify the critical points where pro- and anticapitalist climate politics part company.

Major Klein uses the word "reckless" a lot and rails far more against deregulated capitalism and market fundamentalism than against capitalism or markets. This Klein advocates "dignified work" and valorizes climate action as a "massive job creator" for "good clean jobs."[7] Climate change, here, offers the opportunity to finish the "unfinished business" of civil rights and decolonial struggles at the level of their perceived demand for the right to work. It "could bring the jobs and homes that Martin Luther King dreamed of; it could bring jobs and clean water to Native communities."[8] Unsurprisingly, "the resources for this just transition must ultimately come from the state," as part of a "Marshall Plan for the Earth."[9] This Klein differs from the ideology of neoliberal market fundamentalists in understanding that market mechanisms won't create this transition without decisive state intervention. She therefore calls for the state to "create the market for further investments."[10]

Klein sees that "there is plenty of room to make a profit in a zero-carbon economy, but the profit motive is not going to be the midwife for that great transformation."[11] Given a climate movement that challenges the endless drive for profit as the organizing principle of social life, why limit its ambition to capitalism minus the fossil fuels? Are co-ops and a welfare state just a more "human" face on exploitation? We may be forced to settle for that as a compromise, depending on what

4. Klein, *This Changes Everything*, 118.
5. Klein, *This Changes Everything*, 10 and 157. For more on climate populism, see "Climate Populism and the People's Climate March" in Section IV of this volume.
6. Klein, *This Changes Everything*, 94 and 450.
7. Klein, *This Changes Everything*, 124, 125, 178 (among several other spots).
8. Klein, *This Changes Everything*, 458.
9. Klein, *This Changes Everything*, 5 and 401.
10. Klein, *This Changes Everything*, 401.
11. Klein, *This Changes Everything*, 252.

becomes of the relationship between the state and social forces, not to mention capital and labor. Yet the hard part of even getting that far would be generating powerful and combative movements in the first place. Once that's done, why throw real transformative power aside and stop at a greener shade of Keynes?[12]

Indeed, financing this redistributive program through taxes on fossil fuel profits makes the state dependent on continued burning of fossil fuels. But if the confiscation and redistribution of fossil fuel profits is just a one-off stage in a more radical transformation, this criticism can perhaps be avoided. In other words, the major Klein needs the minor Klein to provide the "People's Shock, a blow from below" in order to transition to a post-fossil fuel embedded liberalism.[13] But does the minor Klein need the major Klein?

Minor Klein sets her sights on "the fundamental imperative at the heart of our economic model: grow or die . . . a drive that goes much deeper than the trade history of the past few decades."[14] Yet major Klein's focus on moral values leads her to the proposition that there are "sectors that are not governed by the drive for increased yearly profit (the public sector, co-ops, local businesses, nonprofits)."[15] Such sectors can't be so neatly disentangled from the desire for growth, as evidenced by her analysis of the global environmental nonprofit organization The Nature Conservancy, who she reveals had in fact bargained with the devil by benefiting from years of oil extraction on a Texas prairie preserve which it had inherited from the gas giant Mobil in 1995. Such "Big Greens," she argues, frequently operate in the interests of the capitalists who make up their Board of Directors, causing significant ecological damage.

Disappointingly, major Klein identifies only one impersonal mechanism by which the drive for profits is enforced: the fiduciary obligation of corporate directors to maximize shareholder returns. This needs to be expanded to map the numerous mechanisms that together overdetermine the drive for endless growth, in the sense of there being multiple, distributed sufficient causes. Without understanding this web of causes it becomes easier to posit the kind of reformist solutions aimed at a single cause, as major Klein endorses.

The mechanisms that seem to require growth can be listed almost endlessly. Investors require returns, Klein notes, as do debtors. Firms need to generate surplus

12. On Klein's "Green Keynesianism," see Geoff Mann and Joel Wainwright, *Climate Leviathan: A Political Theory of Our Planetary Future* (London and New York: Verso Books, 2018), 167–70.

13. Klein, *This Changes Everything*, 10.

14. Klein, *This Changes Everything*, 21.

15. Klein, *This Changes Everything*, 93.

to cover unforeseen costs or losses. Surplus is also required to reinvest in future productivity, which may even reduce short-term profits and dividends. For some, income growth is seen to be a means for addressing climate catastrophe but this can lead to perverse outcomes, such as the series of aforementioned oil wells and profits benefitting The Nature Conservancy. Firms may need to achieve economies of scale through expansion. Finally, the pressure of competition requires a constant search for new arenas of surplus to secure first-mover advantage or to avoid being left behind. States, which have no shareholders, further rely on surplus. GDP underwrites hard military/trade power and soft aid/opinion power, while growth expands the tax base and allows the state to keep rolling over the national debt.

Capitalism is a system of relations which compels certain behaviors regardless of individuals' values. Indeed, for supporters of capitalism, this is precisely the virtue of the invisible hand. The impersonal weave of this system needs to be taken into account when considering whether local businesses, nonprofits, the public sector, and cooperatives can transcend its logic.

TWO POLITICS

A similar tension between the two Kleins emerges in *This Changes Everything*'s treatment of politics. Major Klein sees the necessity to "reclaim our democracies from corrosive corporate influence," "challeng[e] corruption," and "demand (and create political leadership)" capable of "saying no to powerful corporations."[16] This Klein sees the problem as rooted in the malign influence of corporate campaign donations, resulting in "this corroded state of our political systems."[17] She expresses bitter disappointment with Barack Obama's failure to fulfill his climate promises, but hopes that if social movements mobilize, then "politicians interested in reelection won't be able to ignore them forever."[18]

Major Klein seems to accept that the Right, in the name of the free market, are the principle "dismantlers of the state."[19] This is a common but erroneous conception that accepts the public-facing claims of neoliberalism at face value. Like other forms of economic liberalism, neoliberalism has always been a project of state power. There is a reason that the neoliberal era is usually dated to the coming to power of Augusto Pinochet, Margaret Thatcher, and Ronald Reagan. Consequently,

16. Klein, *This Changes Everything*, 7, 119, 152.
17. Klein, *This Changes Everything*, 361.
18. Klein, *This Changes Everything*, 382.
19. Klein, *This Changes Everything*, 52.

while major Klein sets out to reclaim the state for the people, it is not clear to which pre-corroded, pre-corrupted state she refers. The period she seems to have in mind is the populism of the New Deal. Yet she aptly notes that "social movement pressure created the conditions" for such sweeping, redistributive policies.[20] The corporate-capitalist character was still there, but the interests of workers and mass movements were able to force certain compromises.

The minor Klein's politics are resistant to the state's attraction. This Klein acknowledges that "even in countries with enlightened laws as in Bolivia and Ecuador, the state still pushes ahead with extractive projects without the consent of the Indigenous people who rely on those lands."[21] She coolly observes that "the reason industry can get away with this has little to do with what is legal and everything to do with raw political power . . . and anyway, the police are controlled by the state."[22] Here she also advocates a very different form of political organization, replacing the "collusion between corporations and the state" which has reduced communities to "little more than . . . 'waste earth'" with, instead, "new democratic processes, including neighborhood assemblies."[23]

These two understandings of politics are incommensurable. In the practices of social movements, we see nascent mass forms of noncoercive political organization. The dynamics of the urban Aymara social movement in Bolivia and the recent Indigenous "Idle No More" protests in/against the state of Canada suggest the strength of these struggles is inversely proportional to their coupling to the state. The major Klein underestimates the extent to which recognition and representation within the state is one of the techniques deployed by hand-wringing liberals in demobilizing Indigenous, and by extension, migrant, antiracist, feminist, and workers' struggles.

Because they are incommensurable, these two modes of politics cannot coexist for long. This is why we glimpse egalitarian movements from below in occupied squares and workplaces, in self-organized disaster relief efforts, and on the barricades of the nascent climate movement dubbed "Blockadia." Contrary to major Klein, when examining such social formations, we glimpse a world beyond and against the state. The major Klein is critical of such movements' supposed tendency to "structurelessness" over "institution-building," but such a claim mistakes the use of structures and institutions directed against and beyond the state (mutual aid,

20. Klein, *This Changes Everything*, 454.
21. Klein, *This Changes Everything*, 377.
22. Klein, *This Changes Everything*, 378.
23. Klein, *This Changes Everything*, 363 and 406.

assemblies, the Zapatista principle of *mandar obedeciendo* [leading by obeying]) for the absence of structure and institution.[24]

The split between politics through the state and politics against and beyond the state is especially apparent in the question of border politics, given the direct imperative to migration wrought by capitalism. The European Union, often positioned by the major Klein as a force for good in the fight against ecological devastation, has some of the deadliest borders in the world. Developing a mass politics against and beyond the state is vital to prevent domination by xenocidal lifeboat ethics in the future.

REPRODUCTIVE FUTURISM OR REGENERATIVE KINSHIP?

Klein strikes a more inspiring note when she heralds "a new kind of reproductive rights movement, one fighting not only for the reproductive rights of women, but for the reproductive rights of the planet as a whole."[25] Her account of capitalism is nowhere so materialist as where she declaims how "our economic system . . . does not value women's reproductive labor, pays caregivers miserably, teachers almost as badly, and we generally hear about female reproduction only when men are trying to regulate it."[26] Throughout *This Changes Everything*, kids, children, baby dolphins, fertility, future generations, seeds, and the quest for pregnancy poetically and politically structure Klein's arguments.

Much of this, however, is framed by the major Klein. She pits the good life, Mother Earth, New Age health practices, rural tranquility, and a natural drive towards reproduction and regeneration against the non-natures of pollution, technology, cities, Frankenstein, and in vitro fertilization. She rejects the disfigured, monstrous character of the Earth. "We did not create it; it created—and sustains—us. . . . the solution . . . is not to fix the world, it is to fix ourselves."[27]

These themes culminate in the last full chapter of the book, "The Right to Regenerate," in which Klein draws parallels between her own fraught attempts to conceive a child amid the stress of modern urban life with the wider inability to value nature or emulate its gift for life. "As a culture," she claims, "we do a very poor job of protecting, valuing, or even noticing fertility—not just among humans but

---

24. Klein, *This Changes Everything*, 158. *Mandar obedeciendo* is collective decisionmaking principle utilized by the Zapatistas. See John Holloway, "The Concept of Power and the Zapatistas," *Common Sense* 19 (1996): 20–27.
25. Klein, *This Changes Everything*, 443.
26. Klein, *This Changes Everything*, 430.
27. Klein, *This Changes Everything*, 279.

across life's spectrum."[28] This conviction leads her to wonder: "Was it even possible to be a real environmentalist if you didn't have kids?"[29] While many environmentalists in fact fall into a symmetrical yet opposite trap by morally abjuring natalism, Klein's argument espouses reproductive futurism—a troubling equivocation between the production of (proper) children and the prospect for a (proper) future for humanity.

Major Klein's reproductive futurism devalues the queer and the now. It leaves little space for the refusal of reproduction or the discussion of how and what is reproduced. It denies the intrinsic and equal worth of, as minor Klein puts it, "exiles from nature."[30] It leaves the private form of the family largely unquestioned, the essentialism of the term "Mother Earth" untouched, and the primacy of "fertility" intact. It betrays a romanticism that hinges on a "natural" life-domain somehow separate from capitalism.

Such an ontological and moral split is empirically wrong and depoliticizing. Capitalism is not simply a logic of "short-term economic growth" imposed by some (predominantly) middle-aged white men upon a separate, rich biotic world whose fundamental logic is long-term growth, circular regeneration, and life. Indeed, both capital and reactionary thought are often premised on forms of "regeneration." Such rhetoric provides cover, for example, for the razing of public housing, opposition to abortion, and the UN-led "carbon offset" forests that Klein critiques. The latter result in Indigenous people being driven from their homes so that industrial activity elsewhere can be counted as "sustainable." Capitalism is not something antithetical to nature but, as Jason W. Moore argues, a way of organizing it.[31] Nature cannot express, in any unsullied way, what we are fighting for. We cannot simply affirm life. Rather, we must always ask: what forms of life? Which natures? For whom?

Minor Klein appears on the cusp of exploring such a "monstrous" conception of nature by naming a "kinship of the infertile," which we read as solidarity with nonreproductive lifeforms.[32] This openness to the complex desires of the dispossessed—including the desire for consumption, collective luxury, safety, "development," and freedom from "shitty" work—shows an occasional attunement to the already technological, entangled, human-nonhuman character of nature: a

28. Klein, *This Changes Everything*, 430.
29. Klein, *This Changes Everything*, 423.
30. Klein, *This Changes Everything*, 424.
31. Moore, *Capitalism in the Web of Life*. See also "Organizing Nature in the Midst of Crisis" in Section II of this book.
32. Klein, *This Changes Everything*, 427.

cyborg Earth. Such a recognition does not condemn us to geoengineering or other Promethean technofixes, but rather decenters maternity to make room for the "unnatural," the technological, and the nonfertile among the "we" coming into being in the struggles in Blockadia.

Cyborg Earth is not a foregone concession to evil technoscience but a site of struggles over the commons just like any other. A cyborg everything-ism reorients us towards practices that repurpose existing technologies and organizations of nature through bricolage. Minor Klein hints at a more hybrid, anti-austerity sensibility of this kind; one that does not recoil from monstrous entanglements of human, nonhuman, and technological natures. She is doubtful about her desire for pregnancy and implies that if ecological crisis changes everything, surely it changes the institution of the family too. Disappointingly, the priority of incorporating a nonreproductive politics into the "regenerative" struggles of anticapitalism vanishes at the very moment in the narrative when Klein, at last, conceives a baby.

HAS EVERYTHING CHANGED?

Much more could be said about this inspiring and perplexing book. Klein's ability to appeal to both direct-action radicals and conservative journalists at the same time reflects the polyvocal character of *This Changes Everything*—there's something in there for everyone. This reflects, at least in part, Klein's desire for a broad populist politics which unites left and right; one that draws on a social base of small local businesses but forms alliances with Indigenous movements, trade unions, more affluent homeowners, campus activists, NGOs, states, and supranational organizations.

While a text can sustain such dissonance, movements face real tactical and strategic choices. *This Changes Everything* is a rich resource, but one from which the reader needs to pick out certain lines of argument in order to turn them against others.

There is an irreconcilable difference between a politics which seeks to back small local businesses against big global ones and a politics which seeks to challenge whether business of any size is a desirable model of social organization. Likewise, a politics which sees social movements as only providing a potential constituency for electoral campaigns remains locked in and limited by the statist politics of representative democracy. By contrast, the kind of mass social movement practices evident in, for example, contemporary Indigenous struggles, point to a rejection of recognition and representation within the state, which prefigures mass forms of noncoercive political power beyond and against it.[33]

Finally, there is also a contradiction between a regenerative politics understood as reproductive rights writ large and as decentering biological reproduction at just one moment of the multitude of human, nonhuman, and technological natures inextricably entangled in reproducing a world worth living in—and fighting for. The history of reproductive struggles is characterized as much by a refusal of imposed motherhood as by affirmation of fertility.

The New Deal compromise Klein looks to as a precedent required not only powerful social movements, but also institutionalized representatives able to police them. The success of the trade union bureaucracies in turning the sit-down strikes of the thirties into the orderly industrial relations of the fifties also undermined the potency of workers' struggle. The New Deal reforms not only helped stabilize capitalism (and its endless drive for profits), but the compromise helped undermine the disruptive power which had forced the concessions in the first place. The problem with the major Klein's program is therefore not that it "doesn't go far enough" by some radical standard, but that in leaving central capitalist institutions in place, it is ultimately self-defeating from an ecological point of view.

A compromise is always a provisional balance between opposing forces. The major Klein aims to hold at the fulcrum; the minor Klein invites us to push past the social tipping point and see what world our struggles can create on the other side. Major Klein wants to "ride the tiger." Minor Klein hints that tigers need no riders, and don't take too kindly to those who seek to harness them for their own ends. Are today's participants in climate movements willing to put their bodies and lives on the line, only to find that their dreams served to enlist them as foot soldiers for a modest Keynesian agenda? While Klein's major politics point to such a recuperative closure, her minor politics are more open-ended, accommodating more utopian impulses that operate within, against, and beyond the now.

We are tempted to go so far to say that the politics of major Klein are parasitic upon those of minor Klein. Political gains through the state are dependent upon social movements operating outside and or against the state. If "parasitic" might seem strong, it is worth noting that parasites can, of course, provide some benefit to the host. The role played by a small business that provides free space for migrant solidarity meetings should not be overlooked. A state that provides a decent standard of living and foregoes the draconian repression of activism creates more

33. Glen Coulthard, *Red Skin, White Masks*.

space for Blockadia to act in. But if cyborg Earth is to flourish, it must be this latter politics that becomes parasitic upon the former, turning upon its host in a more radical transformation of social relations than the host could ever have provided.

Our perspective leaves us with key questions after reading *This Changes Everything*, and these might be productively framed through the book's title. Does "this" refer to the threat a particular form of capitalism poses to romantic, heteronormative "nature"? Or the imbrication of capital and climate resisted by a cyborgian struggle that goes beyond it? Do the "changes" amount to modest adjustments so that the institutions major Klein favors can persist in a post-carbon world? Or is it the necessary abolitionism minor Klein hints at? Does "everything" refer just to capitalism? Or does it include everything that needs abolishing: the state, work, borders, generalized commodification, prisons, profits, the family, local businesses, settler colonialism, Keynesian economics, and the image of the baby's face as synonymous with "the future"?

Taken at face value, Klein's apparent advocacy of both positions at once reads as contradictory or possibly incoherent. Reading *This Changes Everything* more charitably as an unintended dialogue between a major and a minor Klein allows us to identify tensions in the text that reflect real bifurcations in responding to something as all-encompassing as climate change. The minor lines of thought point beyond the book to connect with texts, concepts, and live debates within social movements. They point beyond the ecocidal logic of endless growth, and further beyond capitalism, the state, and ecological crisis. At various points Klein seems on the cusp of pursuing this minor line, but each time she reprises a major refrain. The distance between a major and a minor key is a single note, but everything depends on the difference.

CLIMATE POPULISM AND THE PEOPLE'S CLIMATE MARCH

First published September 2014

Transnational climate politics has been dominated by the United Nations Framework Convention on Climate Change (UNFCCC) meetings, twenty-five of which have occurred since 1992. Millions of hours have been devoted to negotiation by states of various culpabilities, and the names of these conferences—Kyoto, Copenhagen, Paris—offer only oneupmanship in their complete failure. Climate activism of various political tendencies has 'conference hopped' these events, protesting, blocking, urging more transformative action but confined to patronizingly powerless 'civil society' zones outside the real action. In this context, in 2010 the first World People's Conference on Climate Change and the Rights of Mother Earth was held in Cochabamba, Bolivia as an alternative. The site was chosen to inherit the Cochabambans' years of political struggle

against water privatization, as facilitated by the World Bank and the International Monetary Fund, later understood as the "water wars." Though not·without its shortcomings, the agreement reached by the 30,000 people gathered at Cochabamba was drastically different from any Council of Parties (COP) agenda, and precipitated a transformation in strategy around the world.

This essay was an attempt to think through the potentialities and limits of naming expansively 'the people' as the subject of global climate justice, as was the case at the People's Climate March in 2014, owing something to the Cochabamba event. This we understood as "climate populism," an attempt to build a mass transnational movement unifying disparate strands around the world. Climate populism could be seen in the march itself and assessments therein, in Naomi Klein's *This Changes Everything*, published to coincide with the march (see previous essay), in the quaint calls for 'civil disobedience' in the face of government corruption, oil and gas denialism, and lack of meaningful citizen participation. Our essay was an attempt to intervene in this space, to see if something productive could be made of the movement moment, but like 'climate populism' itself, our analysis was far too limited. We border on a kind of entryism in suggesting climate activists could be desubjectivated from such an event. We fail to take into account the ephemerality of the performative act of street protest and its lack of demands. We shunt the political/representational problems of unity and diversity rather quickly. We failed to recognize how easily such an event would be hijacked by those it opposed—UN Secretary-General Ban Ki-moon himself happily participated. Nowhere would such a movement be capable of anything but the most tentative democratic socialisms—those 'left-populisms' that briefly emerged around this time (Syriza, Podemos) have been (for various reasons) utter failures, while the Latin American left-populisms lie now in extractive ruins, precipitated by the shortcomings of various combinations of capital, the state, and international military and economic intervention.

Occupy Wall Street and the movements of the squares similarly tried to sidestep the problem of class composition and, perhaps too early, asserted their unity under a single banner: the 99%, Indignados. The climate populists continue to run into a similar problem. The working class, as the Endnotes collective argue, is "unified only in separation . . . capital is the unity of the world, and its replacement cannot be just one thing. It will have to be many."[1] At

its core, the too-fast unifying elements of the left-populist project tend to further marginalize the radical difference presented by Black and Indigenous thought, for example, while potentially feeding into the nationalist populisms of the far right around the world.

While we have updated this essay to take such critiques into account and to gesture towards a way in which 'the people' might function to include and promote, rather than suppress, difference, we are not convinced about (and do not fully agree on) our success to this end. In this, perhaps Out of the Woods itself constitutes the collective form we advocate: we have a singular identity, but this does not discount the possibility of disagreement among us.

We work within historical circumstances not of our choosing, and must approach existing tendencies clearly and critically. Aspects of climate populism stumble on in youth climate movements, in the generalized antipolitics of 'saving humanity,' perhaps with only minimal lessons learned from past events. Our analysis here was of a moment, in a way that we hope provides lessons about the limitations of a populist style and moment. The antidote to the vagueness of 'people's movements' is a more explicit, generous, and humble understanding of the subject who might carry forward the horizon of communist politics. The old, reductive understanding of 'the working class' to fill that role is clearly insufficient, with its masculinities, national boundaries, and teleological strategies. Climate populism has been wildly insufficient in producing an expansive enough subject as well, reinforcing a kind of multicultural activism fully compatible with green capitalism. What kind of comradery would it take to uphold the communism that must emerge to confront the scale and breadth of ecological crisis?

---

1. Endnotes Collective, *Endnotes 4: Unity in Separation* (London: Endnotes, 2015), 166 and 167.

In September 2014, the UN Climate Summit was held in New York City. After yet another disappointing round of global talks on climate change failed to produce even the most flimsy of agreements among participating countries, UN Secretary-General Ban Ki-moon invited world leaders of UN member states as well as individuals and groups from finance, business, and "civil society" to "catalyze ambitious action on the ground to reduce emissions and strengthen climate resilience and mobilize political will" in a voluntary meeting external to the United Nations Framework Convention on Climate Change (UNFCCC) process.[1]

As if sprung from the process itself, a large mobilization emerged external to this already external meeting, dubbed the People's Climate March. While the UN Climate Summit is touted as critical to the attempt by political and technocratic elites to (re)affirm capitalist hegemony on our climate futures,[2] the People's Climate March, we argue, must take this critical opportunity to invent a subject—"the people"—already coming into being through today's climate crisis.

POPULISM: A BRIEF DEFINITION

The name "People's Climate March" immediately locates the mobilization within the political tradition of populism—those movements that claim to speak for the people themselves. Populism can be seen as a constitutive force that lies at the heart of the political itself, while the politics of climate change pose some specific problems that highlight both the limits and possibilities of populist strategies. Before turning to climate change, we begin with a more general discussion of populism and political power.

Populist movements are, generally speaking, those which emphasize that "the people" should hold primary decision-making power, and any political or economic force is derived from this collective or community. The people is positioned against an elite, establishment or corrupt outside intruder, the latter serving as the racialized locus for right-populism's xenophobia. Attempts to define populism beyond this schematic, however, are somewhat more difficult. Many historical and contemporary socialist and anarchist movements can be thought of as populist, including the Luddites and Diggers, the Russian and Cuban socialist movements,

---

1. UNFCCC is the international framework under which the Kyoto Protocol was negotiated, and under which annual "Conferences of the Parties" (COPs) are held. COP 26 will be held in Glasgow in November 2020.

2. For a summary of the UNFCCC process and organized opposition, both anticapitalist and less so, see David Ciplet, Mizan R. Khan, and J. Timmons Roberts, *Power in a Warming World: The New Global Politics of Climate Change and the Remaking of Environmental Inequality* (Cambridge, MA: MIT Press, 2015).

and aspects of Occupy Wall Street. What these leftist movements share is a broader commitment to "the sovereignty of the people." But the language of populism is certainly capable of fitting right-wing agendas, even though it is not reducible to them.[3]

The People's Party active in turn-of-the-century United States founded itself on Jeffersonian yeoman morality while opposing governmental and corporate elitism. Agriculturalist movements from South Asia to South America are often founded on "the land of the people" versus urban techno-elites or international financiers. Tea Party movements in the US and UKIP in the UK and elsewhere in Europe, with their distrust of cultural elites, governmental institutions, and immigrants also clearly rely on populist tropes, as do populist movements more radically invested in a charismatic leader like Juan Perón in 1970s Argentina, or in more recent South American Left leaders, such as Hugo Chavez in Venezuela and Evo Morales in Bolivia. Extended far enough in this direction, populist movements can easily fall into dictatorship or fascism. Similarly troubling is the plasticity of populist movements to fit cross-class interests, often resulting in mere reforms and the capture of working-class political action to reify the status quo.[4]

We believe that Ernesto Laclau is on to something when he writes that "the construction of the people is the political act par excellence," and is thus not so much a type of movement, but a political logic that pervades the social itself.[5] Following Laclau's argument in *On Populist Reason*, we can say that no populist political movement exists without an attempt to take the demands of some people and extend them to a universal level. On the other hand, in claiming to speak for *all* people, an expansive populism can suffer from an inability to locate what is excluded from its political vision. Political populism thus reaches towards a universal while manifesting specific exclusions from this sphere.

For Laclau, the dangers of populism outlined above are in some sense secondary to the political logic itself. The form or tendency that a populist movement actualizes depends entirely on the relations of resonance it is able to create with other movements—and what other demands, positions, or relationships are sacrificed

---

3. In the time since we originally wrote this piece in 2014, populism as a signifier is increasingly used by an embattled political liberalism to dismiss political projects beyond apolitical technocracy and to equate political projects right and left that do the above. See Thea Riofrancos, "Democracy Without the People," *n+1* (blog), February 6, 2017, https://nplusonemag.com/online-only/online-only/democracy-without-the-people/.

4. Battlescarred, "Pedlars of Reformism and the Occupy Movement," *libcom.org* (blog), September 13, 2012, http://libcom.org/library/pedlars-reformism-occupy-movement. This piece discusses cross-class populism in Occupy, particularly alliances with "Main Street USA" businesses. The apocryphal Joe Hill quote speaks to this concern: "the people and the working class have nothing in common."

5. Ernesto Laclau, *On Populist Reason* (London and New York: Verso Books, 2005), 154.

in this attempt.[6] For example, if class interests are sacrificed to create a cross-class alliance, a movement will attain a different character and composition while perhaps losing more radical possibilities. More prevalent in the United States recently have been anarchist-libertarian resonances, which often subordinate anticapitalist demands to a distrust of hierarchy.

"THE PEOPLE ARE MISSING"

Many left thinkers are of course skeptical of the vagaries of populism and for good reason. However, critiques (especially from Leninists) often rely on a belief that "the people" can't possibly organize themselves in a political manner without being members of a party, submitting to an agenda, or investing in a leader. Like liberals and social democrats, Leninists can appear afraid of the people, seen as either an unorganized aggregation of individuals or an irrational mass or crowd. These fears we feel are rooted as much in liberal—and even more so, neoliberal—demands that we view only individuals as sovereign and capable of political decision making. By contrast, a potential benefit of (particular forms of) populism is the redirection of our attention to the power of collective decision-making and the necessity of antagonism.

Finally, populism exposes the construction of political relations over a mythic or essentialized "people." If populist movements display a collective desire to make decisions as a collective, then we must dispel with the founding myth of Western populism: that the people are a self-contained, organic, and pre-existing entity. Instead, we wonder whether there could be a populism that makes its own foundational myth present (and contestable). This is against most populism, which (by assumed or actual necessity) hides the artifice of "the people." Populism sometimes celebrates the moment of foundation or constitution, but as an act that brought something always-already extant/potent into political power. Such Jeffersonian logic begins from the representational premise that one can speak of and for the people immediately and directly. The United States Declaration of Independence begins with "We the People. . ." and, along with other texts central to the founding of the United States has, somewhat confusingly, continued to serve

---

6. Here we might think of the manner in which abolitionist positions vis-à-vis the police, prisons, and borders have (since this piece was first written) been abandoned, downplayed or watered down by sections of the Left who have allied themselves to the statist projects of Jeremy Corbyn and Bernie Sanders; both of whom frequently use populist rhetoric of speaking for "the people." Such statist shifts result in the further exclusion of poor, racialized subjects; prisoners; and migrants from the category of "the people."

as inspiration for both American left theorists and in everyday left-leaning protests in the United States.[7]

Instead, we follow Gilles Deleuze's imperative: today, "the people are missing," and thus must be invented.[8] Deleuze looks to the cinema of Third World to theorize the creation of "the people" in the cramped spaces of survival in the midst of oppression. Here we need to posit an important break with "the people" called into being as part of the foundation of the United States. As George Ciccariello-Maher notes in his reading of Enrique Dussel and Venezuelan struggle, in Latin America, the *pueblo* is not a subject of "cold unity, but dynamic combat."[9] This "people" can thus have an antagonistic relationship with the state and collectively produces itself. Such a political movement might be consonant with "the prefiguration of the people who are missing," as Deleuze puts it.[10] The always-ongoing collective act of self-invention is thus the creation of a space of resonance in which heterogeneous groups are gathered to create the language necessary to invent a people (and thus, as a corollary, exclude certain others from such a collective).

In order to clarify, let's take Occupy Wall Street as a recent and familiar example of a quasi-populist movement. Although much was made of Occupy's lack of demands, the slogans and memes developed by the movement reproduce some common populist tropes. For this group, the slogan "we are the 99 percent" functioned as a common signifier that bound the group together and named its foundational people, even as these "people" argued over what this meant. Crucially, the function of "we are the 99 percent" was not simply unity and consolidation around a fragile "we," but to name those who are excluded from that we: the 1 percent. In this sense, Occupy had a clear political logic, even if elsewhere its demands may have seemed vague or empty.[11]

The problems and dangers of populist political logics abound. Who or what is excluded? How are they to be differentiated from those included? Are these exclusions based on a foundational understanding of racial or national characteristics, even thinly disguised? Who decides? Are leaders elected or emergent? These more general questions are familiar to any participants in radical liberatory social

---

7. See, for example: Anneke Campbell and Thomas Linzey, *We the People: Stories from the Community Rights Movement in the United States* (Oakland: PM Press, 2016).

8. Gilles Deleuze, *Cinema 2: The Time Image*, trans. Hugh Tomlinson and Robert Galeta (Minneapolis: University of Minnesota Press, 1989), 216.

9. Ciccariello-Maher, *Decolonizing Dialectics*, 129.

10. Deleuze, *Cinema 2*, 224.

11. See, for example: Dean, *The Communist Horizon*; Joe Lowndes and Dorian Warren, "Occupy Wall Street: A Twenty-First Century Populist Movement?," *Dissent Magazine*, October 21, 2011, https://www.dissentmagazine.org/online_articles/occupy-wall-street-a-twenty-first-century-populist-movement.

movements, and are outside the scope of this brief essay. Instead, we'd like to turn to the particular problems that a climate populism faces today.

CLIMATE POPULISM

The above analysis should sufficiently reveal the populist tendencies of the PCM and other global climate movements. However, given how populism functions as a political logic, there are three specific problems facing climate populism: a scalar politics above/below the national; lack of a clearly defined enemy; and an inability to name itself as anticapitalist.

Climate politics, like the antiglobalization movements before it, has defined itself by moving away from national scale analysis and towards both localism and transnationalism. Populist movements in affluent parts of North America and Europe especially rely on localism and (bio)regionalism in a manner that has tendencies towards reactionary ecology. On the other hand, a newer and increasingly prevalent trend has been to depoliticize climate by claiming that it affects everyone (humanity is at stake) thus subsuming smaller "political" demands.

The People's Climate March has faced criticism for the inclusion of "Green Zionist" groups in its heterogeneous base. Supporters of the PCM have deflected criticism by claiming that climate change is a "bigger issue" than Palestinian liberation; that if we don't stop climate change the Middle East will be uninhabitable; that the whole endeavor is bringing a political agenda to what was supposed to be a unifying, broad coalition. It is all too easy to imagine future mass mobilizations in support of overtly fascist and nationalist climate policies: anti-immigration, hoarding national energy resources, against reproductive rights, or in favor more drastic measures like forced sterilization. Unlike, for example, socialist populist movements, which named a clear enemy in the bourgeoisie, North American climate populism has struggled to define its opponents. Currently, environmental and environmental justice organizations have defined "the fossil fuel industry," "large corporations," and "climate deniers" as the enemy. But opposing corporate greed ignores the root of the problem, which is the specifically capitalist form of resource extraction.

Simply opposing climate denialists does very little but promote a consensus belief that climate change is "real" while ignoring the divisive political aspect of our responses or creative transformations.[12] Repurposing our movements towards fossil

---

12. For this reason, Alain Badiou has forsaken ecological politics because "everything which is consensual is

fuel abolitionism requires a more strategic and international focus on the social relations that entrench fossil fuels as the basis of our economies and political formations.

Finally, of course, the PCM has yet to take a specifically liberatory, communist—or frankly, *any* avowed political position aside from a vague hope for a redemptive future. Outlining steps towards climate justice and solidarity with Indigenous nations are incredibly important but can have the effect of ringing hollow when many in the PCM are still committed to green capitalism, overt nationalism and "energy independence," and urban climate resilience (let alone allowing the Zionism mentioned above). 350.org leader and admittedly "reluctant" US environmental movement leader Bill McKibben often makes statements like, "There's nothing radical about what we're doing here. We're just Americans, interested in preserving a country and a planet that looks and feels something like the ones we were born on."[13]

Given the above, why should communists and anticapitalists invest anything in the PCM and similar populist movements?

INVENTING CLIMATE PEOPLE?

The "people" of the People's Climate Movement are still missing, and the reluctance to name a demand creates an opportunity for a politics to be created and seized in the "cramped space" of New York City still experiencing the aftereffects of Hurricane Sandy. There has been some disagreement among the Green Left on whether and how to participate or relate to groups like the People's Climate March.[14] On the one hand, the gathering clearly takes messaging tools and financial resources from establishment environmental groups like 350.org that position its appeal to a wider (admittedly American) population. Engaging with the UN process and its overt reliance on financial and green capitalist sustainable development seems to reaffirm that the power to decide our climate futures is in the hands of those very institutions that put us in this problem. However, the Green Left may be shirking opportunities for tactical engagement when it avoids events like the PCM. The call to "Flood Wall Street" in conjunction with the PCM has the ability to redirect the force of the movement away from the UN Climate Summit and towards finance capital.

without a doubt bad for human emancipation." [Alain Badiou, *Alain Badiou: Live Theory*, ed. Oliver Feltham (London and New York: Bloomsbury Academic, 2008), 139.]

13. Bill McKibben, *Oil and Honey: The Education of an Unlikely Activist* (New York: Times Books, 2013), 44.

14. See the essays collected at Rising Tide North America, "Growing the Roots to Weather the Storm: Reflections on the Global Movement for Climate Justice," *Growing the Roots to Weather the Storm*, 2014, http://growingdeeproots.risingtidenorthamerica.org/?cat=6.

Who constitutes "the people" of the PCM is inconsistent and open ended. Its qualification and uptake is missing (strategically, one would assume). The bourgeois politics of climate change has mired itself between the likely reformist and green capitalist UNFCCC process and a geopolitical debate over coming resource wars. Because climate populism has been understood by the UN process ambivalently, as merely a constitutive force for official politics, it is likely to be less relevant for the sanctioned political order. This situation will produce increasing frustration that can be channeled into collective, creative action, as we saw with the 2010 World People's Forum on Climate Change and the Rights of Mother Earth. Such a response is in excess of, and against, the vague reformism of the PCM.

Similar to the logic of repurposing as bricolage—which we have discussed as a characteristic of disaster communism—we cannot simply negate the form, content or techniques of existing social movements. While the PCM in no way espouses a politics we'd like, critique here seems to simply produce a chauvinistic melancholia and fatalism. These movements too must be infiltrated, repurposed, and redirected towards an alternative definition of "the people."

Other sympathetic groups are involved in the PCM, including blocs like Free Palestine and a number of more clandestinely organized anticapitalist blocs. But the greatest opportunity of the PCM might be its vast potential for desubjectivication of climate activists away from UNFCCC, 350.org, and similar institutions attempting to recuperate the mobilization, and towards liberatory, inventive, and collectively anticapitalist social formations. While our underground movements are important right now, don't underestimate the importance of capturing a collectivizing name for the people—one not for everyone, but for those already marginalized by climate change and attempts to govern it, a name that specifically excludes capitalist futures and one around which we can constitute and consolidate our desires.

APRÈS MOI LE DÉLUGE!

FOSSIL FUEL ABOLITIONISM
AND THE CARBON BUBBLE

First published November 2017

This essay has two clear ambitions. The first is to provide a critical understanding of the centrality of fossil fuels to social life, incorporating both the infrastructure that extracts them and the economic system that dictates this extractive process. Our success in fulfilling this ambition is largely thanks to our use of Marx's thought, which provides both structuring definition and analytical incision. Marx's careful reading of history lends the essay a comprehensive analysis of capital, infrastructure and value, which cuts through the complexity to provide a clear vision of the problem. It is our use of Marx's methods which also leads us to realize a fundamental impasse: a vast amount of capital is caught up in a form of extraction that ensures further devastation of living and nonliving things. Mitigating this devastation requires the abandonment of this capital.

The need to find a way out of this impasse provides the impetus for the second ambition of this essay: to identify a historical precendent for such a mass abandonment of capital. To do so, we turn to the abolition of slavery in the United States. Here, however the essay seems to lose its analytical incision

and clarity. Whereas the first part of the essay is based on Marx's critique, the second part lacks structure, and the argument drifts into generalization and imprecision. On reflection, what is so striking about our analysis of abolition is the absence of the critique that enabled and enacted abolition in the first place: the thought and practice of the Black radical tradition. Our argument should be oriented by thinkers such as W.E.B. Du Bois, and works such as his Black Reconstruction. *Instead of our imprecise account of different abolitionist tendencies, Du Bois would push us to begin with the fact that abolition was the result of a 'general strike' of enslaved Black people. Where the essay looks to the abolition of slavery to prove that a mass destruction of capital has theoretical precedent, Du Bois forces us to consider the material practices that made abolition possible. Within and against the ongoing catastrophe of enslavement, in which value ensured devestation, Black people organized for a different world. If we are to properly think through abolition in our work, this is the beginning we must return to.*

To stand any chance of keeping global warming below dangerous levels, a large percentage of fossil fuel reserves need to stay in the ground unburned.[1] A figure as mainstream as Nicholas Stern estimates that sixty to eighty percent of current (i.e., "known") reserves are unburnable.[2] Energy transitions have occurred in the past: "from peat and charcoal (1450s–1830s), to coal (1750s–1950s), to oil and natural gas (1870s–present)."[3]

These transitions were additive, while a transition to renewables must be eliminative. This means leaving resources in the ground worth hundreds of billions of dollars. The amount that Stern advocates leaving in the ground supports share values of around $4 trillion, constitutes approximately twenty-seven percent of US Gross National Income ($15 trillion), and around five percent of Gross World Product ($85 trillion).[4] Infrastructures that depend on fossil fuels—refineries, manufacturing processes, power plants, airports, cars, conventional agriculture—must also be factored into the equation.

It doesn't stop there. Capital is not just a lump of money; it is a lump of money *in motion*. Money begets money. Writing off the current value of already invested wealth would involve writing off all future wealth that capitalists in the petrochemical industries hope to own as a result of their investments, not to mention that of other capitalists selling products to petrochemical companies. Capitalists will not write off this wealth voluntarily out of ecological concern. As ever, they hope to cash in before calamity strikes.

"*Après moi le déluge!* [after me, the flood] is the watchword of every capitalist and of every capitalist nation," Karl Marx argued.[5] Marx's environmental metaphor now carries a valence he could hardly have foreseen. A huge amount of capital is bound up with fossil fuels, creating a strong path dependency for capitalist development. Nevertheless, some discussion of unburnable fossil fuels has begun to enter policy-making circles, framed around the notion of a "carbon bubble." To

1. This assumes no dramatic breakthroughs in carbon capture and storage (CCS) technology, but even then a huge new CCS infrastructure would need to be rolled out at great cost.
2. Carbon Tracker, "Wasted Capital and Stranded Assets," Carbon Tracker Initiative, April 19, 2013, https://www.carbontracker.org/reports/unburnable-carbon-wasted-capital-and-stranded-assets/.
3. Jason W. Moore, "Ecology, Capital, and the Nature of Our Times: Accumulation and Crisis in the Capitalist World-Ecology," *Journal of World-Systems Research* 17, no. 1 (2011): 128.
4. Bill McKibben, "Global Warming's Terrifying New Math," *Rolling Stone*, July 19, 2012, https://www.rolling-stone.com/politics/politics-news/global-warmings-terrifying-new-math-188550/. The eagle-eyed may notice we're comparing a stock (asset values) with a flow (annual national income/world product). The reason for this will become apparent, as it allows the most meaningful comparison across century timescales by relating asset values to the contemporary size of the economy without complicated attempts to take account of inflation.
5. Marx, *Capital* (1976), 381.

meet the internationally agreed warming limit of 2°C (or a "safer" limit of 1.5°C), large amounts of fossil fuels must go unburned. But firms' stock prices are valued on the basis those reserves *can* be burned. Therefore, these firms are artificially overvalued: there's a carbon bubble, much like the subprime bubble, and just as the latter did it is at risk of bursting. A recent editorial in the journal *Nature Climate Change* puts it in the following terms:

> By consistently overvaluing the fossil fuel assets of companies, the argument goes, the world's financial markets are busily inflating a "carbon bubble" which, if burst, could spell ruin for investors. It is no surprise then that individuals, corporations and pension fund holders are beginning to wake up to the risk and either starting to divest from fossil fuels or seriously considering it. Even the World Bank has stopped lending for new coal-fired power plants.[6]

The most obvious objection here is that while the subprime bubble arose out of endogenous crisis tendencies within capitalism, any mechanism to leave fossil fuels in the ground would be politically determined. That is, it would be undertaken voluntarily by states; either unilaterally, or as part of a multilateral agreement such as the supposed Kyoto Protocol successor which is meant to be agreed at the United Nations Climate Change Conference in Paris in 2015. In other words, the carbon bubble only becomes a bubble if policymakers (the states themselves, given the UN's limited power) legislate it such. It's hard to see why states would create such an economic crunch voluntarily.[7]

6. Nature Climate Change, "Carbon Bubble Toil and Trouble," *Nature Climate Change* 4, no. 4 (April 2014): 229. Available at https://doi.org/10.1038/nclimate2193.

7. Indeed, they did not. The "Paris Agreement," a successor Agreement due to come into force in 2020, does not require that *any* fossil fuels are left in the ground (in fact, it does not once mention the term "fossil fuels"). Rather, the 185 parties to the agreement (184 states and the EU; to be reduced to 183 states in November 2020 when the US' withdrawal from the treaty goes into effect) must establish how they will contribute to the Agreement's overall goal of preventing a rise in global average temperature of more than 2°C above pre-industrial levels. Limiting fossil fuel extraction has been suggested as one key way states may do this, but given the Agreement does not legally bind states to specific targets the chances of success seem slim. The Stockholm Environment Institute, for example, suggests reductions in fossil fuel extraction as a key method states should consider using, but their constructivist approach places a naive faith in the power of environmental "norms" to spread through global and domestic civil society against the desires of capital. [See "The Paris Agreement | UNFCCC," accessed April 26, 2019, https://unfccc.int/process-and-meetings/the-paris-agreement/the-paris-agreement; Georgia Piggot et al., *Addressing Fossil Fuel Production Under the UNFCCC: Paris and Beyond* (Seattle, WA: Stockholm Environment Institute, 2017).]

LEAVE IT IN THE GROUND!

A widely vaunted proposal for leaving fossil fuels in the ground came from the democratic socialist government of Rafael Correa in Ecuador in 2007. They proposed that in return for leaving the oil reserves under its Yasuní National Park untouched (a move that would also have protected the Indigenous Tagaeri and Taromenane populations), the world should pay Ecuador half of the estimated value of the reserves up-front. This proposal was positively received by mainstream environmentalists as well as Marxists and Indigenous activists.[8] The rest of the world, not so much: "Spain chipped in a couple million. So did the Andean Development Bank and the Inter-American Development Bank. The UN and other private individuals raised some funds. But in the end, Ecuador only raised $13 million, a far cry from the $3.6 billion Correa had sought."[9] It later transpired that during the proposal, Ecuador had been secretly negotiating extraction concessions with China.[10] Whether or not this was evidence of duplicity, or simply pragmatic contingency planning, the failure of the Yasuní proposal seriously damaged the prospects for leaving fossil fuel reserves unexploited.[11]

If states aren't willing to pay to leave resources in the ground, they're certainly not going to just write them off. The passage from Marx quoted earlier continues: "capital is reckless of the health or length of life of the laborer, *unless under compulsion from society*."[12] Marx was discussing struggles over the length of the working day in Victorian-era England, but the argument applies just as pertinently to the health of the ecosystem. Indeed, the chapter begins with another famous ecological analogy: "Capital asks no questions about the length of life of labor-power. What interests it is purely and simply the maximum of labor-power that can be

---

8. For a mainstream view endorsed by the Prince of Wales, see Mike Berners-Lee, Duncan Clark, and Bill McKibben, *The Burning Question: We Can't Burn Half the World's Oil, Coal, and Gas. So How Do We Quit?* (Vancouver: Greystone Books, 2013). For a Marxist perspective, see John Bellamy Foster, Richard York, and Brett Clark, *The Ecological Rift: Capitalism's War on the Earth* (New York: Monthly Review Press, 2011).

9. Brad Plumer, "Ecuador Asked the World to Pay It Not to Drill for Oil. The World Said No," *Washington Post*, August 16, 2013, https://www.washingtonpost.com/news/wonk/wp/2013/08/16/ecuador-asked-the-world-to-pay-it-not-to-drill-for-oil-the-world-said-no/.

10. David Hill, "Ecuador Pursued China Oil Deal While Pledging to Protect Yasuní, Papers Show," *The Guardian*, February 19, 2014, sec. Environment, https://www.theguardian.com/environment/2014/feb/19/ecuador-oil-china-yasuni.

11. Drilling for oil by Ecuador's state oil company in the park began in early 2018, prompting concerns about global carbon-emission increases and more ecological and colonial devastation within the park itself. It has been overseen by Correa's former Vice President, Lenín Moreno, who has been praised for seeming to entertain environmental concerns. This is evidence that "better" politicians and decent "norms" are insufficient in the face of capital's thirst for fossil fuels. [See Jonathan Watts, "New Round of Oil Drilling Goes Deeper into Ecuador's Yasuní National Park," *The Guardian*, January 10, 2018, sec. Environment, https://www.theguardian.com/environment/2018/jan/10/new-round-of-oil-drilling-goes-deeper-into-ecuadors-yasuni-national-park.]

12. Marx, *Capital* (1976), 381. [Emphasis added.]

set in motion in a working day. It attains this objective by shortening the life of labor-power, *in the same way* as a greedy farmer snatches more produce from the soil by robbing it of its fertility."[13] We do not believe that today's imperial powers (supranational organizations, states, and policymakers) will take the necessary steps to provide "compulsion from society" to mitigate the exhaustion of the planet. States and supranational organizations (where imperial powers generally remain the most powerful actors) may be able to wield their power towards the ends of climate justice, but only under extraordinary amounts of external compulsion.

The Aufheben collective argue that when it comes to climate change there are two principal capitalist factions. The first group are "fossil capitalists," corporate denizens like the Koch Brothers who fund the climate change denial public relations industry.[14] "The second group, "green capitalists," are spearheaded by sections of finance capital, especially the reinsurance industry. The former's interests are bound with the reproduction of ecological crisis; the latter's with mitigating some of its effects.[15] It is important to recognize these two factions, and note that their material interests diverge somewhat when it comes to climate policy. Given the value of assets that would have to be written off and developmental path dependency (such as the potentially stranded assets of fossil fuel infrastructure like pipelines), it's hard to see how green capitalists could gain ascendency, except perhaps by harnessing social movements into a limited reformist version of a Green New Deal at the expense of the fossil capitalists.[16]

13. Marx, *Capital* (1976), 376.

14. Naomi Oreskes and Erik M. Conway, *Merchants of Doubt: How a Handful of Scientists Obscured the Truth on Issues from Tobacco Smoke to Global Warming* (New York: Bloomsbury Publishing USA, 2011).

15. Aufheben, "The Climate Crisis . . . and the New Green Capitalism?," *Libcom.org* (blog), 2014, http://libcom. org/library/climate-crisis-...and-new-green-capitalism. Reinsurance is insurance for insurance companies. They allow insurers to spread risk such that if a catastrophic event leads to a high number of insurance payouts (increasingly likely as symptoms of ecological crisis such as flooding, fires, rising sea levels, and coastal erosion become more regular occurrences) the insurers will themselves be covered. Thus, they have financial interests in reducing some of the effects of climate change.

It is important to note, too, that the interests of these factions aren't always so straightforward. The threats from climate change may allow reinsurers to charge higher premiums for their services. For example, the world's highest-grossing reinsurance company, Munich Reinsurance Company (or Munich Re Group), describes climate change as providing "risks and opportunities." Shell has publicly called for a one-trillion-ton cap on greenhouse gas emissions. This could be pure greenwashing PR (as private lobbying the other way continues) or, it could be enlightened self-interest if they think they're well-positioned with renewable technology and intellectual property. They could also think a hard emissions limit will drive the necessary state support for Carbon Capture and Storage, thus protecting the value of their fossil fuel assets.

16. We have already used the qualifier "some" in talking of green capitalism's desire to mitigate the effects of climate change, and the limit to any such program are a corollary of this. Reinsurers, for example, only have incentive to mitigate the effects of climate change that people take out policies to cover. Typically, unpaid laborers, gestators, prisoners, stateless peoples, slum dwellers, peasants, and isolated Indigenous peoples (overlapping categories, of course) do not take out insurance, and so there is no incentive for reinsurers to mitigate or ward off ecological crisis as it affects them. And save where they play a role in the reproduction of capital, "environmental" phenomena such as wildlife diversity, and soil, water and air quality, are unlikely to be of interest to reinsurers.

A move from the current fossil fuel–dependent development path to a sustainable alternative requires a significant period of economic contraction, stagnation, and restructuring. There are few historical precedents for such a rapid and dramatic reorganization outside of wartime. Perhaps the closest historical parallel is with the abolition of slavery in the United States. There are, of course, major historical differences between nineteenth-century and present-day capitalism. While we don't want to collapse these, we do think it's worth looking at previous instances where large amounts of capital have been written off due to a change of development path.[17]

ABOLITIONISM IN THE NINETEENTH-CENTURY UNITED STATES

In 1850, the US population was about twenty-three million, of whom over three million—thirteen percent—were enslaved people. According to Thomas Piketty, US national capital in 1850 was 440 percent of national income, with enslaved people valued at 108 percent of national income.[18] In today's terms, the value of enslaved people would therefore equate to assets of around $16 trillion. In addition to this "present value" (that is the value of assets at the time, rather than with future potential earnings factored in), abolition had wider costs, devaluing investments in labor processes and entire industries that used slave labor, which were at the center of the United States economy, including textile mills in the North. The value of this broader infrastructure is almost impossible to estimate with any degree of accuracy. Mitigating the effects of climate change would similarly destroy huge amounts of both present value and potential future worth. It would also require intense social conflict similar to that aroused by the abolition of slavery. Though it is important not to retrospectively map contemporary ideological and material positions onto abolitionists, we can see a similar range of political approaches and material relations to climate change in the contemporary world.

In their motivations, material relationship to slavery, and preferred tactics and strategies, abolitionists in the nineteenth-century United States constituted a broad political spectrum. They consisted of enslaved people, free Black people (many of them former slaves), and white allies. Some sought an immediate end to slavery, some saw abolition as a necessarily or preferably gradual process. Some advocated

17. One important difference is that in 1850, cheap and abundant fossil fuels could be used to substitute for human muscle power once slavery was abolished, and indeed continuously, as rising productivity replaced labor with machines. The era of cheap energy is now over.
18. See Piketty's numbers at piketty.pse.ens.fr/files/capital21c/en/pdf/supp/TS4.2.pdf.

punishment for slave owners. Others sought compensation for the loss of their "property" (sometimes the full value, sometimes a percentage). Some abolitionists saw womens' participation in the movement as necessary to its success, and as a parallel, though not directly analogous struggle. Others insisted that women be excluded from the movement. Some advocated direct action, violent and nonviolent, and including the act of winning (one's own) freedom. Some preferred the use of the ballot box and legal methods. Some saw abolition as an act of loyalty to the founding principles of the United States. The most radical sought a dissolution of the Union. Some wanted freed slaves to live as American citizens. Others felt they should be deported to Africa. Some were clear and consistent in their abolitionism, while others changed over time. Many abolitionists held seemingly contradictory positions at the same time, such as owning slavery-adjacent industries. Many recognized the necessary and messy intersections between divergent (and sometimes opposed) approaches.

The history and political economy of slavery, abolition, and the United States Civil War is of course complicated and contentious. Nonetheless, we can assert that slaveholders and the "ruling-class" of political forces had a kind of uneasy coexistence that grew increasingly troubled over the nineteenth century. The US federal government kept those forces in relative balance within the political economy of the time. As the United States expanded territorially, essentially in the fashion of a settler colonialism, conflicts broke out over where slavery would and would not be permitted.

Colonial expansion occurred in the context of a growth market in cotton and other slave-made goods and rising prices in enslaved people. Which is to say, there was a great deal of money to be made by investing in slavery and putting enslaved people to work. That great deal of money to be made was getting bigger over time, creating pressures for the further expansion of slavery. To point this out is not to reduce chattel slavery to a series of self-consistent economic motivations that could be ascribed solely to capitalism. Rather, much like global environmental injustice today, the political economy of slavery coincided with and relied upon series of ideological and libidinal relations of power and domination. These were sometimes in concert, sometimes in contradiction with the aims of capital.

The conflicts that emerged from such contradictions occurred with greater frequency over time and became more intense. Slave owners and other ruling-class forces were increasingly polarized. That polarization and the nature of the conflicts that occurred were dramatically shaped by abolitionists, who pressed upon the preexisting and growing tensions between ruling-class factions. Some of these tensions arose from policy and economic changes, but abolitionists also created new tensions and politicized the nature of slavery itself, coding slavery as a moral wrong (though as we noted above, abolitionists differed among themselves and their differences included what kind of wrong they said slavery was).

FOSSIL FUEL ABOLITIONISM?

The point of re-examining slavery abolitionism is to lay out a rough analogy with the present. This should not occlude the ways in which the roots of the climate crisis can be found in the same systems of economic and agricultural organization under the height of chattel slavery. Many of our present ecological crises are *extensions* of these same struggles over who and what counts as a human.[19] In calling attention to these parallels and continuities, we hope to examine contemporary struggles with greater sharpness and clarity, to suss out the tensions that might end up most important in a new cycle of struggles with the inheritance of the past.

Today too, we see policy and economic developments creating tensions among different capitalist-class factions. "The capitalist class" has been deeply fractured, perhaps irreversibly so, by the 2008 financial crisis and is split on whether the climate crisis provides an opportunity for profit or not. Among tendencies opposed to the climate crisis one can find moral crusaders, insurrectionary types (romantic and pragmatic), those invested in meaningful direct action (e.g., blockades and riots), as well as those more interested in civil disobedience. Other axes of the contemporary field of climate and ecological politics include profound disagreement about scale, nature, inequality, and universality. For our part, we maintain a principled skepticism about the very possibility that the dangers of ecological crisis can be avoided without a meaningful overthrow of capitalism. If climate change is to be controlled while capitalism continues, we do not believe such a transformation could emerge from capitalists and their governments in the absence of militant social movements. As residents of this planet, we would of course welcome the control of our contribution to the climate over the runaway climate change that is becoming a terrifying new possibility. Yet we are not just residents. We want more from the future than simply "continuing to live on this planet." We are libertarian communists who want humanity to live on this planet *in particular ways*. It is clear that not all who are opposed to the climate crisis will be forces for genuine human

19. See, for example, Fred Moten, "Blackness and Poetry," *ARCADE*, no. 55 (2015), https://arcade.stanford.edu/content/blackness-and-poetry-0.

emancipation, any more than all abolitionists were forces for genuine emancipation. How can the response to this crisis be pushed towards more expansive ends? How do we ensure we work against ecological crisis (as a crisis unevenly affecting humans) rather than just climate change (as a walling off of one specific form of crisis from broader socio-environmental phenomena)? How will we proceed?

None of this is to say that civil war is in the cards (or to look romantically upon a concept of "civil war"). Nor is it to accept that the Civil War in the United States was fought simply to abolish slavery (as opposed to emancipation being a useful ploy to destabilize the South, for example). Instead, abolition demonstrates how the emergence of a social movement can tip the balance of forces between rival developmental trajectories within the capitalist class. This is especially true when the material interests concerned are too divergent, due to path dependency, to be reconciled by normal capitalist processes of responding to the general will (e.g., party politics). It's worth noting that by 1880, US national capital had recovered to 422 percent of national income, with enslaved people now accounting for zero percent (according to Piketty's data), which suggests other forms of capital quickly grew to make up for the loss.

MASS STRUGGLE, NOT ENLIGHTENED POLICY

Climate change seems likely only be averted/contained in two scenarios: a libertarian-communist revolution overthrowing states and capital and instituting a global commons (which would address other elements of the ecological crisis); or a powerful, militant movement that emerges around climate change harnessed by intercapitalist and interstate conflict, analogous to the American Civil War. These aren't entirely distinct, as green capitalists would surely attempt to harness any nascent communist movement to restructure capital in their interests, "reforming to preserve, not to overthrow."[20]

If a mass anticlimate change movement emerges, the Left could mistake it for being revolutionary, as many of us have done in the past. It's important therefore to understand the extent to which environmentalism can oppose particular factions of capital, or even the current path of capitalist development, without necessarily

20. British Prime Minister Earl Grey, discussing suffrage in 1831. [Quoted in Solidarity Federation, *Fighting for Ourselves: Anarcho-Syndicalism and the Class Struggle*, October 27, 2012, 73. Available at https://libcom.org/library/fighting-ourselves-anarcho-syndicalism-class-struggle-solidarity-federation.] See also Mario Tronti regarding how capital develops through incorporating elements of the struggle of the working class, turning it against those who struggle. [Mario Tronti, *Workers and Capital* (London and New York: Verso, 2019).]

opposing the capital relation itself.[21] As Newell and Paterson warned in *Climate Capitalism* (2010), "many of the union activists in the thirties wanted to abolish capitalism, but in practice contributed to a better-regulated and more successful version of it."[22]

It's also important to recognize that any movement powerful enough to force big reforms has the potential to overflow the constraints of the capital relation and its social forms (states, private property, commodities, wage labor, etc.). People and movements are dynamic. A struggle that begins as reformist in its demands and social vision may come to demand more thoroughgoing change: organizing against climate change may expand to encompass ecological crisis more broadly. Movements will test the limits of already existing institutions and, in effect, learn lessons in the process.[23]

In noting the possibility for climate justice movements to take on a reformist character we are not advocating abstention from them. Instead, we are rather advocating the need for communists to push for those movements to expand their political vision to a more expansive sense of liberation. We should not assume that climate justice movements will be inherently or explicitly anticapitalist, but they will need to challenge the current organization of capitalism if they are to succeed. In doing so, they open up the space for a more fundamental socio-ecological transformation.

In the absence of such a movement, the dynamic equilibrium between recuperation and rupture may seem moot. There is, after all, precious little to recuperate. But if the business-as-usual trajectory is to be reversed, the dynamics of social movements and intercapitalist wrangling will be crucial. The value of the assets and infrastructure that need to be written off to mitigate dangerous climate change is staggering and of a similar order of magnitude to slavery.[24] A

21. Militant reformism is a term that has come up in (and really, developed through) conversation among people on libcom's blogs. The discussion has spread across multiple blog posts and is hard to summarize. For a representative sample, see s.nappalos, "Responding to the Growing Importance of the State in the Workers' Movement," *Libcom.org* (blog), February 4, 2014, http://libcom.org/blog/responding-growing-importance-state-workers-movement-04022014.

22. Peter Newell and Matthew Paterson, *Climate Capitalism: Global Warming and the Transformation of the Global Economy* (Cambridge: Cambridge University Press, 2010), 180.

23. This is, in some ways, the story of antipipeline movements in North America. In this case, some participants in the movement began as Not-In-My-Back-Yard (NIMBY) opponents, many of whom were members of the reactionary "Tea Party" movement. Although the tendencies moved in several directions, some of these same individuals and groups would work in alliance with Indigenous peoples at the Standing Rock blockade a decade later, demanding no less than a radical decolonization.

24. It is worth noting that the write-off in the United States was a social and economic earthquake. And this earthquake involved massive military casualties. Petrochemical companies like Shell are already notorious for their support for brutal dictatorships, most infamously with their collusion in the execution of Nigerian activist Ken Saro-Wiwa (Shell paid $15.5 million in an out-of-court settlement to his family to avoid admitting

write-off of this magnitude won't happen without a push, but we shouldn't underestimate the capacity of the capitalist system to harness social movements in order to transform and preserve itself. Attempts to create and burst a carbon bubble through top-down policy channels are a dead end. Such a massive economic write-off will not be voluntarily undertaken given the likely impact on stock markets (down) and food prices (up) alone.[25] In this sense, responses to climate change, whether reforms or revolution, are in the hands of social movements. There are many potential flashpoints for such a movement: food riots, movements for free transport, anti-extraction and anti-airport struggles. Such struggles seek to block the current path of development, with its intensive use of fossil fuels, and are vital for that reason. Their emergence may also generate a crisis which opens up the possibility of radical social transformation.

liability). Recently this has been dramatized in the political video game *Oligarchy*. These behaviors have happened in a period of relative profitability for the petrocapitalists. Once we start talking about actual destruction of their capital, they may well be willing to resort to much more widespread use of violence to maintain the existence of their industry. Here too, contemporary anti-pipeline movements are instructive.

25. See Joshua Clover on 2018's *gilets jaunes* [yellow vests] protest as "an early example of an approaching wave of climate riots." [Joshua Clover, "The Roundabout Riots," *Versobooks.com* (blog), December 9, 2018, https://www.versobooks.com/blogs/4161-the-roundabout-riots.]

DISASTER COMMUNISM

THE USES OF DISASTER

First published October 2018

Just as it felt fitting that the BASE Magazine interview should open the book, it feels very right that "Disaster Communism" should form its close. If the interview captures us throwing new ideas into the air, then this essay marks the first time we were really able to pull them together and fix them to the ground. "The Uses of Disaster" is thus the best articulation of how we think now. It is the product of all the pieces that came before it, both the ones included in this book and those that have been left out. These essays, however, do not represent some form of linear advance that leads, inexorably, up to our present form of thinking. If they did, it would have made sense to republish them chronologically.

Instead, the contents of this book reflect the reality of working together. There was no straighforward progress from there to here, but rather an uncertain passage, replete with missteps, set-backs, and dead-ends. Writing these introductions has reminded us of many failures, of the many times we felt trapped in seemingly inescapable impasses. And yet, together, we have always found a way to improvise an escape, to take what we need from the wreckage and build something new. Writing with each other, like all forms of social life, holds the promise of bricolage towards something else; of getting out of the woods, together.

"What is this feeling that crops up during so many disasters?," Rebecca Solnit asks in *A Paradise Built in Hell*.[1] Examining human responses to earthquakes, fires, explosions, terrorist attacks, and hurricanes over the last century, Solnit asserts that the commonplace idea that disasters reveal the worst in human nature is misguided. Instead, she argues, such events reveal "an emotion graver than happiness but still positive." This purposeful hope galvanizes what she calls "disaster communities."[2] When the prevailing social order temporarily fails, a host of "extraordinary communities" constituted by acts of mutual aid spring up in response. Solnit's examples include Hurricane Katrina, 9/11, and the Mexico City earthquake of 1985, but these are not exceptions to the rule. Most contemporary disasters feature at least brief fleeting moments in which we forget social differences and help each other. Alas, when the disaster passes, these communities seem to subside. In the terms of Solnit's *A Paradise*, the "great contemporary task" we face is the prevention of that subsidence, "the recovery of this closeness and purpose without crisis or pressure."[3] Given the calamity of our warming planet, this task becomes ever more urgent. How do we dismantle the social orders that make disasters so disastrous, while at the same time making the extraordinary human behavior they elicit ordinary?

Solnit's argument rings true even if one is less optimistic than her about the inherent value of "community." In the hells of the present, we find the tools we need to build other worlds as well as tantalizing glimpses of possibilities often thought unachievable. This is not prefiguration, nor cause for celebration or even optimism. But it is cause for hope.

For this hope to be realized, however, we must go beyond Solnit's empirical focus on how communities respond to specific disaster-events and grasp the character of the capitalist disaster. This is not simply a series of punctuated dates and place names—Katrina, Harvey, and Irma, 1755, 1906, and 1985—but an ongoing condition. For many, the ordinary is a disaster. Any coherent response to such continuous *ordinary disaster* will likewise have to be widespread and durable in order to succeed. Building paradise in hell is not enough: we must work *against* hell and go *beyond* it. We need more than disaster communities. We need disaster communism.[4]

1. Solnit, *A Paradise Built in Hell*, 5.
2. Solnit, *A Paradise Built in Hell*, 5.
3. Solnit, *A Paradise Built in Hell*, 113.
4. Revolts Now, "Disaster Communism & Anarchy in the Streets," *Revolts Now* (blog), April 10, 2011, https://revoltsnow.wordpress.com/2011/04/10/166/. Out of the Woods' 2014 three-part exploration of disaster communism can be found at libcom.org. Part I discusses the spontaneous communities of mutual aid typically formed in disaster situations. [Out of the Woods, "Disaster Communism Part 1 – Disaster Communities," *Libcom.org* (blog), May 8, 2014, libcom.org/blog/disaster-communism-part-1-disaster-communities-08052014]. Part II shifts

Assuredly, in calling for disaster communism, we are not suggesting that the occurrence of more and more frequent ecosocial nightmares will somehow inevitably produce ever riper conditions for communism. We cannot adopt the perverse fatalism of "the worse, the better" nor wait for some final hurricane to blow away the old order. Rather, we are noting that even the largest scale and most terrifying of these extraordinary disasters can interrupt the ordinary disaster that is, most of the time, too large to fully comprehend. These are moments of interruption that, while horrific for human life, might also spell disaster *for* capitalism.

Disaster communism is not a brand-new type of politics divorced from existing struggles. Rather, it is a revolutionary process of developing our collective capacity to endure and flourish that emerges from these struggles. Disaster communism is a movement within, against, and beyond ongoing capitalist disaster. It seeks to address how the numerous projects creating mini-paradises in hell might cohere into something more than ephemeral communities. It adds a clarifying epithet to the already ongoing political projects that pit themselves against the state and capital, and notes how such projects overflow their bounds. It orients the movement of a collective power that, rendered palpable during extraordinary disasters, was there all along. It was there especially in places and among groups who have experienced ordinary disaster for hundreds of years. Ecological crisis brings the skills central to those struggles into focus.

DISASTER CAPITALISM, CAPITAL AS DISASTER

Geographer Neil Smith makes a convincing argument that there's no such thing as a natural disaster.[5] Naming disasters "natural" occludes the fact that they are just as much the product of political and social divisions as they are of climatic or geological forces. If an earthquake destroys the poorly constructed and badly maintained low-income housing in a town but leaves the well-built homes of the rich standing, blaming nature lets states, developers, and slumlords off the hook (not to mention

---

to a wider angle, considering the possibility of communism in a world soon to be, and perhaps already, committed to climate chaos. [Out of the Woods, "Disaster Communism Part 2 – Communisation and Concrete Utopia," *Libcom.org* (blog), May 14, 2014, http://libcom.org/blog/disaster-communism-part-2-communisation-concrete-utopia-14052014]. Part III seeks to pull the micro-moments of disaster communities and the macro-problematics of disaster communization together through engagement with a debate over logistics. [Out of the Woods, "Disaster Communism Part 3 – Logistics, Repurposing, Bricolage," *Libcom.org* (blog), May 22, 2014, http://libcom.org/blog/disaster-communism-part-3-logistics-repurposing-bricolage-22052014]. While it may be helpful for readers to refer back to these earlier debates (especially if interested in communization theory or counter-logistics), the updated version of our argument, published here, offers a more refined and complete analysis.

5. Smith, "There's No Such Thing as a Natural Disaster."

the capitalist economy that produces such inequalities in the first place). Disasters are always coproductions in which natural forces such as plate tectonics and weather systems work together with social, political, and economic forces.

The ways in which extraordinary disasters play out, then, cannot be separated from the ordinary disaster conditions in which they occur. The Category 4 Hurricane Maria that devastated the US colony of Puerto Rico, leaving residents without fresh water, was a disastrous event. But to begin the narrative there obscures the fact that, prior to the hurricane, "99.5 percent of Puerto Rico's population was served by community water systems in violation of the Safe Drinking Water Act," while "69.4 percent of people on the island were served by water sources that violated SDWA's health standards."[6] Nor should such devastating events eclipse slower moving disasters, such as in Flint, Michigan, where decades of neglect and the industrial pollution of the surrounding Flint River and Great Lakes combined with new austerity policies to strategically abandon working-class, majority Black and Latinx communities without clean water. Easily overlooked because they lack the spectacular power of a hurricane or earthquake, such drawn-out disasters blur the boundary between disaster-as-event and disaster-as-condition. What comes as a sudden and unexpected jolt for many is a matter of intensified everyday reality for others.[7]

Climate change significantly increases the frequency and severity of both slow- and fast-moving disasters. Global warming means an increased amount of energy circulating in the atmosphere and ocean surfaces. For example, when warm oceans create low air-pressure cells, thermal energy, under the influence of the Earth's rotation, is converted into the kinetic energy characteristic of swirling hurricanes and tropical storms. Warmer temperatures give rise to more energy, which has to somehow be expressed. Energy can only change form; it cannot be destroyed.

The physics of this are fiendishly complex and difficult to model. Yet it is still possible to make forecasts. The latest report from the UN's Intergovernmental Panel on Climate Change suggests climate change will disrupt food and water supplies. It will damage homes and infrastructure and bring droughts and floods, heat waves and hurricanes, storm surges and wildfires.[8] Advances in attribution science and knowledge gleaned from the single-degree Celsius of global warming already in effect make it possible to quantify the contribution of climate change to individual

6. Natural Resources Defense Council, "Threats on Tap: Drinking Water Violations in Puerto Rico," NRDC, May 2, 2017, https://www.nrdc.org/resources/threats-tap-drinking-water-violations-puerto-rico.

7. Rob Nixon, *Slow Violence and the Environmentalism of the Poor* (Cambridge, MA: Harvard University Press, 2011).

8. Intergovernmental Panel on Climate Change, "Global Warming of 1.5°C," special report for UN Environment Programme (UNEP) 2018 Annual Report, https://www.ipcc.ch/sr15/.

extreme weather events. We can now link global warming to calamities such as the European heat wave of 2003 and the Russian heat wave of 2010, which both killed tens of thousands of people. This is not to mention countless storms, floods, and other weather events. That climate change is itself human-made (or rather, capitalism-made) further underscores the impossibility of separating disastrous events from disastrous conditions. The relationship between the two runs both ways. Conditions give rise to events which, in turn, further entrench conditions.

The goal of the nation-state during and in the immediate aftermath of extraordinary disasters has usually been to impose order rather than to assist survivors. For this reason, disaster events generally exacerbate the underlying disaster of everyday life under capitalism. Shortly after the 1906 San Francisco earthquake, the army was sent in. Between 50 and 500 survivors were killed, and self-organized search, rescue, and firefighting efforts were disrupted. The state's attempts to manage disaster were a disordering force, destroying bottom-up forms of self-organization. A similar repressive focus on "looters" (i.e., survivors) marked the state's response to Hurricane Katrina in New Orleans. On September 4, 2005 on Danziger Bridge, seven police officers opened fire on a group of Black people attempting to flee the flooded city, killing two and seriously injuring four more. The murder of Black survivors seeking safety neatly illustrates the means by which the state seeks to foreclose emancipatory possibilities that might appear during such disasters.

White supremacy is enabled by mainstream media, which combines dramatic footage of infrastructural and (supposed) societal collapse with Hobbesian narratives of a war of all against all. We have all heard the solemn incantations of news reporters depicting poor, racialized survivors of disasters as "overwhelming" those good, kind, (usually white) aid workers, for example. Footage of ruinous infrastructural and (supposed) societal collapse produces disastrous events as spectacular drama for the news viewer, ensuring the advertising money rolls in.[9] NGOs and aid agencies often repeat such claims. John O'Shea, Director of the Irish medical charity Goal, told *The Guardian* that a US in thrall to "political correctness" was failing to lead and coordinate rescue efforts following the Haitian earthquake of January 2010, resulting in unnecessary deaths.[10]

9. For a critical account of the New Zealand mainstream media's reporting on the Christchurch earthquake of 2010, see Jared, "Amongst the Rubble: A Look at the Christchurch Earthquake from the Bottom Up," *libcom. org* (blog), September 16, 2010, http://libcom.org/news/amongst-rubble-look-christchurch-earthquake-bottom-16092010.

10. Chris McGreal and Esther Addley, "Haiti Aid Agencies Warn: Chaotic and Confusing Relief Effort Is Costing Lives," *The Guardian*, January 18, 2010, sec. World news, https://www.theguardian.com/world/2010/jan/18/haiti-aid-distribution-confusion-warning.

O'Shea's claims were rejected by others on the ground at the time.[11] The actions and understandings described above further run contrary to the recommendations of even the most mainstream sociologists of disaster. In the sixties, for example, Charles Fritz argued trenchantly that the stereotype of widespread antisocial individualism and aggression flourishing during disasters was not grounded in reality:

> Disaster victims rarely exhibit hysterical behavior; a kind of shock-stun behavior is a more common initial response. Even under the worst disaster conditions, people maintain or quickly regain self control and become concerned about the welfare of others. Most of the initial search, rescue, and relief activities are undertaken by disaster victims before the arrival of organized outside aid. Reports of looting in disasters are grossly exaggerated; rates of theft and burglary actually decline in disasters; and much more is given away than stolen. Other forms of antisocial behavior, such as aggression toward others and scapegoating, are rare or nonexistent. Instead, most disasters produce a great increase in social solidarity among the stricken populace, and this newly created solidarity tends to reduce the incidence of most forms of personal and social pathology.[12]

This description of post-disaster organizing appears fundamentally at odds with the reversion to some kind of Hobbesian chaos as suggested by hegemonic ideology. Fritz goes on to note that distinction between disasters and normality can "conveniently overlook the many sources of stress, strain, conflict, and dissatisfaction that are imbedded [sic] in the nature of everyday life."[13] Extraordinary disasters are used to prolong, renew, and extend the ordinary disasters of austerity, privatization, militarization, policing, and borders.[14] This is what Naomi Klein refers to as "disaster capitalism": a vicious cycle in which ordinary disaster conditions exacerbate extraordinary disaster events, in turn intensifying the original conditions.

Disastrous events allow the state to implement what Klein calls "the shock doctrine": redeveloping destroyed housing, energy, and distribution infrastructures to neoliberal standards; pricing the poor out of electricity and clean water; forcing them to relocate to locations even more vulnerable to climate change; and incarcerating them when they resist or try to cross borders to escape this

---

11. Inigo Gilmore, "The Myth of Haiti's Lawless Streets," *The Guardian*, January 20, 2010, sec. Opinion, https://www.theguardian.com/commentisfree/2010/jan/20/haiti-aid-agency-security.

12. Charles E. Fritz, "Disasters and Mental Health: Therapeutic Principles Drawn From Disaster Studies," *Historical and Comparative Disaster Series*, Disaster Research Center, 1996 [1961], 10. Available at http://udspace.udel.edu/handle/19716/1325.

13. Fritz, "Disasters and Mental Health," 22.

14. See Naomi Klein, *The Battle For Paradise: Puerto Rico Takes on the Disaster Capitalists* (Chicago: Haymarket Books, 2018); Miller, *Storming the Wall*; Yarimar Bonilla and Marisol LeBrón, *Aftershocks of Disaster: Puerto Rico Before and After the Storm* (Chicago: Haymarket Books, 2019).

untenable situation.[15] In the aftermath of Hurricane Maria, Puerto Rico has experienced further privatization, worsening labor conditions, and the arrival of green colonialists: supposed do-gooders such as Elon Musk cloaking their latest hypercapitalist ventures behind the thinnest veneer of environmental recovery. The story in Flint is similar too, right down to Musk's offers to "solve" its infrastructural issues.

Forces acting in the interests of state and capital come in a number of shapes, of course. Activists from the Common Ground Collective providing emergency relief in the aftermath of Hurricane Katrina were intensively harassed not only by racist cops but also by armed white locals who seized the opportunity to roleplay a post-apocalyptic, end-times scenario—with the tacit approval and sometimes active facilitation—of the police.[16] Unions, too, have sometimes worked against mutual aid. Following the 1906 San Francisco earthquake, the union-backed Japanese and Korean Exclusion League claimed that Japanese people would feel "at home" in a land of earthquakes, and so, extra effort would be required in order to ensure that California remained "a white man's country."[17] We should not be so naive as to think that such ostensibly racist responses to disaster, in the name of workers' rights, have been confined to the past.

SURVIVAL PENDING REVOLUTION:
THE IMPORTANCE OF SOCIAL REPRODUCTION

What the study of disastrous events and disastrous conditions teaches us is that climate change is not simply an unintended consequence of capitalist production but a crisis of *social reproduction*. The latter is term referring to the self-perpetuating social structures that enable daily and generational survival while at the same time maintaining inequality.[18] Acknowledging this does not just give us a new angle on the problem, however, but also points to a source of hope. It is important to remember that the lives of the poor, the dispossessed, and the colonized are not only shaped by disaster but involve, at every turn, acts of survival and persistence, often in the form of knowledges and skills passed from generation to generation. As the

15. Naomi Klein, *The Shock Doctrine: The Rise of Disaster Capitalism* (New York: Picador, 2007).

16. scott crow, *Black Flags and Windmills: Hope, Anarchy, and the Common Ground Collective* (Oakland: PM Press, 2011).

17. Japanese and Korean Exclusion League, "Asiatic Coolie Invasion," *Organized Labor*, 1906, http://www.sf-museum.net/1906.2/invasion.html.

18. See, for example: Silvia Federici, *Revolution at Point Zero*; Tithi Bhattacharya (ed.), *Social Reproduction Theory: Remapping Class, Recentring Oppression* (London: Pluto Press, 2017); Lewis, *Full Surrogacy Now*.

Potawatomi philosopher Kyle Powys Whyte insists, while Indigenous peoples are quite familiar with disaster in the form of hundreds of years of attempted colonial domination, over those hundreds of years they have developed the skills to resist and survive ordinary and extraordinary disasters.[19]

Autonomist-Marxist feminist scholar Silvia Federici, meanwhile, has shown how capitalism has long endeavored, unsuccessfully, to violently eradicate all forms of noncapitalist survival. She argues that "if the destruction of our means of subsistence is indispensable for the survival of capitalist relations, this must be our terrain of struggle."[20] Such struggle occurred in the aftermath of the 1985 Mexico City earthquake, when landowners and real estate speculators saw an opportunity to finally evict people they'd been wanting to get rid of for a long time. Their attempts to tear down housing that was providing low rental returns and replace it with expensive high-rise condos is a clear example of disaster capitalism at work. However, in this case, working-class residents fought back with great success. Thousands of tenants marched on the presidential palace, demanding the government take possession of damaged homes from their propertied landlords for eventual sale back to their tenants. In response, some 7,000 properties were seized.

Extraordinary disasters create space for the state and capital to entrench their power but also for resistance to these very forms: a "shock doctrine of the Left," to adapt Graham Jones' phrase.[21] The ordinary disaster that is capitalism can in fact be interrupted by these incidents that, while horrific for human life, spell momentary disaster *for* capitalism. In a 1988 analysis of the aforementioned Mexico City disaster titled "The Uses of an Earthquake," Harry Cleaver suggests this is particularly likely with the breakdown of administrative capacity and government authority that follows extraordinary disasters, something that is perhaps even more likely in places where the government relies on surveillance, smart data, and information technologies.[22]

Cleaver also notes the importance of established collective organizing in neighborhoods affected by the earthquake. Survivors in Mexico City had organizational links, a culture of mutual aid, and expectations of solidarity. Tenants knew

19. Kyle Powys Whyte, "Our Ancestors' Dystopia Now: Indigenous Conservation and the Anthropocene," *Routledge Companion to the Environmental Humanities*, eds. Ursula K. Heise, Jon Christensen, and Michelle Niemann (New York: Routledge, 2016), 206–15.

20. Federici, *Revolution at Point Zero*, 89.

21. Graham Jones, *The Shock Doctrine of the Left* (Hoboken, NJ: John Wiley & Sons, 2018). Such a tactic should not be seen as directly analogous to capitalist shock doctrinism, however, because that intentionally *creates* disasters rather than just seeking to utilize them for constructing new worlds. We cannot emphasize enough that we do not hope for disaster as condition or event for communal horizons.

22. Harry Cleaver, "The Uses of an Earthquake," *Midnight Notes* 9 (May 1988): 10–14.

that they had each others' backs because of their past relationships with each other. This point is crucial, for it allows us to understand disaster community not simply as a spontaneous response to extraordinary disasters but rather as the coming to the fore of everyday struggles for survival and subterranean practices of mutual aid. The experience of autonomous organizing against ordinary disaster left residents well equipped to deal with an extraordinary disaster.

Indeed, preexisting relations of support have been most effective in sustaining communities in the wake of Hurricane Maria. Centros de Apoyo Mutuo (CAMs)/ Centers of Mutual Support, a decentralized mutual-aid network drawing on established groups, centers, and practices in post-Maria Puerto Rico, distributed food, cleaned up debris, and rebuilt the island's infrastructure. They have done so more quickly and much more to the needs of residents than networks of international aid. Through the art of bricolage, such mutual-aid programs demonstrate that nonspecialists can quickly pick up and share tools and skills for survival. In a special issue of *The Funambulist* devoted to "Proletarian Fortresses," Marisol LeBrón and Javier Arbona argue: "in building this form of power-with, mutual aid groups further refuse the narrative that only US intervention can save Puerto Rico. They reject the promises of citizenship's protection for the lie it has always been in Puerto Rico, and instead emphasize '*que es el pueblo que va salvar el pueblo*' [it is the people who will save the people]."[23]

Reflecting on solidarity with CAMs, Ricchi, a member of the US-based network Mutual Aid Disaster Relief, suggests that "[t]hose storms have swept by, and they've destroyed many things. . . . By knocking out the energy grid, and cutting access to food and water, they left the island of Borikén [the Indigenous Taíno name for Puerto Rico] dark. But in that darkness countless Boricuas have awoke, and they stay awake late and get up early again, doing the work of reproducing life."[24] This life isn't just mundane, either. Groups organize parties, dancing lessons, and collective cookery sessions so that communal horizons might open beyond despair. Indeed, such actions are not superficial to the life of the commune, but actively demonstrate that "every cook can govern."[25] Altogether, these actions go beyond survival to create new forms of solidarity and collective life.

23. Marisol LeBrón and Javier Arbona, "Resisting Debt and Colonial Disaster in Post-Maria Puerto Rico," *The Funambulist* 16 (March–April 2018). [Requires subscription.]

24. Ricchi, "Dreaming With Our Hands: On Autonomy, In(ter)dependence, and the Regaining of the Commons," *Mutual Aid Disaster Relief* (blog), January 9, 2018, https://mutualaiddisasterrelief.org/dreaming-with-our-hands-on-autonomy-interdependence-and-the-regaining-of-the-commons/.

25. C.L.R. James, "Every Cook Can Govern," *Correspondence*, June 1956, https://www.marxists.org/archive/james-clr/works/1956/06/every-cook.htm. James' reference is to Lenin, but it is his account we reference here.

In a conventional and narrowly economic sense, scarcity exists in these situations. But this scarcity is challenged by an abundance of social links. Extraordinary disasters can push us to recognize that scarcity is a social relation rather than a simple fact of numbers. The way goods and resources are distributed and held determines who can use them as well as their numbers. They show us that we should not be too hasty or certain in associating ecological crisis with increased scarcity.

PARADISE AGAINST HELL

In the aftermath of Hurricane Sandy, a "scarcity" of tools was overcome not through the production or acquisition of more, but through new organization.[26] Tool libraries were set up as alternatives to the individualized, commodified social relations that dominate capitalist society. However, we need more than microcosms for new social relations, not least because such experiments can be valuable for capitalism too. It's important to note that capitalism is not homogenous. What is good for some capitalists is bad for others, and what is bad for individual capitalists over a short period of time may be good for the totality of capitalists in the long run. Thus, while disaster communities might spell bad news for some capitalists and state actors, others will look to mine them for value. The US Department of Homeland Security praised the anarchist-influenced relief efforts of Occupy Sandy after the 2012 hurricane swept through New York.[27] In doing so well what state and market forces could not do, Occupy Sandy kept social life going, giving state and market forces something to recapture once status quo was restored. And they did so at no direct cost to the state.

Such an account is troubling but also partial, of course, and misses the pedagogic value of disaster communities. At its most powerful this is simultaneously negative as well as positive. The resounding "yes" to those deeply felt other ways of living also screams "no" to the ordinary disasters of capital. The social reproduction fostered is a change of direction. It is an attempt to reproduce ourselves *otherwise*, and to resist a return to business-as-usual that leaves our bodies and our ecosystem exhausted.

---

26. "Occupy Sandy Staten Island Tool Library | Occupy Sandy Recovery," accessed April 26, 2019, http://occupysandy.net/?projects=occupy-sandy-staten-island-tool-library.

27. Thomas Hintze, "Homeland Security Study Praises Occupy Sandy, With Murky Intentions," *Truthout*, April 2, 2014, https://truthout.org/articles/dhs-study-praises-occupy-sandy-with-murky-intentions/. See also Ashley Dawson, *Extreme Cities: The Peril and Promise of Urban Life in the Age of Climate Change* (London and New York: Verso Books, 2017), 253. We might also cite the libertarian magazine *Reason*'s frequent celebrations of what it understands as "spontaneous" self-help following disasters, arguing that they are sufficient to absolve the state of any responsibility. Their point, presumably, is that these constitute a proto-market response. This, of course, utterly misunderstands behaviors we understand as mutual aid, while their emphasis on spontaneity ignores the importance of prior relations of solidarity. [See https://reason.com/tags/natural-disasters.]

We see this clearly in many of the disaster communities that spring up in response to borders. As Harsha Walia so brilliantly demonstrates, these do not simply help people mitigate the violence of the border. Instead, they resist the very concept of the border itself, as succinctly put in the widely heard demand for No Borders.[28] Indeed, this very phrase conjures up the simultaneous affirmation and negation we insist upon. Denying a facet of this world while describing features of the next, such a slogan is an operation against and beyond, as well as in, hell.

Such negation will undoubtedly need to go beyond the coziness associated with most understandings of community. When faced with racist cops and vigilantes in the aftermath of Hurricane Katrina, the Common Ground Collective (CGC) engaged in armed self-defense inspired by the Black Panthers and other radical groups.[29] Nor will the conflicts only exist externally. The CGC also had to deal with supporters who seemed more interested in catastrophe tourism than relief efforts. Disaster communities will not be free from the violences that constitute everyday disaster. These include misogyny, white supremacy and saviorism, classism, ableism, and antiblackness. Numerous intersecting forms of oppression will, sadly, leak into their organizing. Disaster communists will thus have to learn how to resolve things *otherwise*, by mobilizing social tools and accountability processes many activists are already developing today.

PARADISE BEYOND HELL

Capitalism is comfortable with community. The term is used to label the resilience capitalism needs to survive ordinary and extraordinary disaster. "Community" is collectivity stripped of all transformative power. Nevertheless, we cannot abandon the concept of community altogether. Such a proposal would be unhelpfully idealistic given its widespread use. But to refer to disaster communities—such as those discussed above—as merely "communities" is to deny their potential, binding them to a present in which they are forever admirable but never transformative. And so, we insist on communism.

Where communism is frequently premised on the material abundance created by capitalist production, disaster communism is grounded in the collective abundance of disaster communities. It is a seizing of the means of social reproduction. We cannot expect, of course, that each and every outcome is immediately

28. Walia, *Undoing Border Imperialism*.
29. crow, *Black Flags and Windmills*.

communist. Private property was not abolished in 1985 by the autonomous community of Tepito in Mexico City, for example. Rather, our use of the term signals the extensive ambition of a movement beyond specific manifestations and outcomes, spread across space and existing beyond extraordinary disasters. It names the ambition to ground nothing less than the whole world in the abundance found in the social reproduction of the disaster community. As such, it fulfills the definition of communism that Marx and Engels give us: "the *real* movement, which abolishes the present state of things."[30]

The communism of disaster communism, then, is a transgressive and transformative mobilization without which the unfolding catastrophe of ecological crisis cannot and will not be stopped. It is simultaneously an undoing of the manifold, structural injustices which perpetuate and draw strength from disaster *and* an enactment of the widespread collective capacity to endure and flourish on a rapidly changing planet. This is an operation *within* that is pitched *against* and opens up space *beyond*. It is hugely ambitious, requiring redistribution of resources at several scales, reparations for colonialism and slavery, expropriation of landed private property for Indigenous peoples, the abolition of fossil fuels, and other monumental projects. We are clearly not there yet. But as Ernst Bloch noted in *The Principle of Hope*, that "not-yet" is also in our present: in the collective responses to disaster, we find that many of the tools for constructing that new world already exist.[31]

When Solnit talks of that emotion "graver than happiness" which animates people in the wake of disaster, she catches "a glimpse of who else we ourselves may be and what else our society could become."[32] Amidst the ruins, within the terrible opening of the interruption, pitched against the conditions that produce and seek to capitalize upon that interruption, we are close to complete change, to the generalization of the knowledge that everything and everyone might yet be transformed. In other words, in the collective response to disaster, we can glimpse a real movement which could yet abolish the present state of things.

---

30. Karl Marx and Friedrich Engels, The *German Ideology: Part One, with Selections from Parts Two and Three*, ed. C. J. Arthur (New York: International Publishers, 1970), 57.

31. Ernst Bloch, *The Principle of Hope, Volume 1*.

32. Solnit, *A Paradise Built in Hell*, 9.

BIBLIOGRAPHY

Abunimah, Ali. "Israeli Lawmaker's Call for Genocide of Palestinians Gets Thousands of Facebook Likes." *The Electronic Intifada* (blog), July 7, 2014. https://electronicintifada.net/blogs/ali-abunimah/israeli-lawmakers-call-genocide-palestinians-gets-thousands-facebook-likes.

Adamczak, Bini. *Communism for Kids.* Translated by Jacob Blumenfeld and Sophie Lewis. Cambridge, MA: MIT Press, 2017.

Alasdair. "Autonomisation, Financialisation, Neoliberalism." *Libcom.org* (blog), January 7, 2013. http://libcom.org/blog/autonomisation-financialisation-neoliberalism-07012013.

Amadiume, Ifi. *Male Daughters, Female Husbands: Gender and Sex in an African Society.* London: Zed Books, 2015.

Anderson, William C., and Zoé Samudzi. *As Black as Resistance: Finding the Conditions for Liberation.* Chico, CA: AK Press, 2018.

Angus, Ian. "The Myth of the Tragedy of the Commons," *Monthly Review Online*, August 25, 2008. https://mronline.org/2008/08/25/the-myth-of-the-tragedy-of-the-commons/.

Anidjar, Gil. *The Jew, the Arab: A History of the Enemy.* Stanford: Stanford University Press, 2003.

Anonymous Contributor. "Occupation, Revolt, Power: The 1st Month of #OccupyICEPHL." *It's Going Down* (blog), August 14, 2018. https://itsgoingdown.org/occupation-revolt-power-the-1st-month-of-occupyicephl/.

_____. "This Movement Is Not Ours, It's Everybody's." *It's Going Down* (blog), July 25, 2018. https://itsgoingdown.org/this-movement-is-not-ours-its-everybodys/.

Arvin, Maile, Eve Tuck, and Angie Morrill. "Decolonizing Feminism: Challenging Connections between Settler Colonialism and Heteropatriarchy." *Feminist Formations*, 2013: 8–34.

Asafu-Adjaye, John, L. Blomquist, Stewart Brand, B. W. Brook, Ruth DeFries, Erle Ellis, Christopher Foreman, D. Keith, M. Lewis, and M. Lynas. "An Ecomodernist Manifesto," 2015. http://www.ecomodernism.org/manifesto-english.

Askew, Kelly, and Rie Odgaard. "Deeds and Misdeeds: Land Titling and Women's Rights in Tanzania." *New Left Review*, no. 118 (2019): 68–85.

Aufheben. "The Climate Crisis . . . and the New Green Capitalism?," *Libcom.org* (blog), 2014. http://libcom.org/library/climate-crisis-...and-new-green-capitalism.

Automnia. "Ecstasy & Warmth." *The Occupied Times* (blog), August 20, 2015. https://theoccupiedtimes.org/?p=14010.

Baccolini, Raffaella, and Tom Moylan. "Conclusion: Critical Dystopia and Possibilities." In *Dark Horizons: Science Fiction and the Dystopian Imagination*, edited by Tom Moylan and Raffaella Baccolini, 233–50. New York and London: Routledge, 2003.

Badgley, Catherine, Jeremy Moghtader, Eileen Quintero, Emily Zakem, M. Jahi Chappell, Katia Aviles-Vazquez, Andrea Samulon, and Ivette Perfecto. "Organic Agriculture and the Global Food Supply." *Renewable Agriculture and Food Systems* 22, no. 2 (2007): 86–108.

Badiou, Alain. *Alain Badiou: Live Theory.* Edited by Oliver Feltham. London and New York: Bloomsbury, 2008.

Bakshi, Sandeep. "Decoloniality, Queerness, and Giddha." In *Decolonizing Sexualities: Transnational Perspectives, Critical Interventions*, edited by Sandeep Bakshi, Suhraiya Jivraj, and Silvia Posocco, 81–99. Oxford: CounterPress, 2016.

Barnes, Hannah. "How Many Climate Migrants Will There Be?" *BBC News*, September 2, 2013. https://www.bbc.com/news/magazine-23899195.

Battistoni, Alyssa. "Monstrous, Duplicated, Potent." *n+1*, April 12, 2017. https://nplusonemag.com/issue-28/reviews/monstrous-duplicated-potent/.

battlescarred. "Pedlars of Reformism and the Occupy Movement." *Libcom.org* (blog), September 13, 2012. http://libcom.org/library/pedlars-reformism-occupy-movement.

Baumard, Maryline. "Give Me Your Tired, Your Poor . . . the Europeans Embracing Migrants." *The Guardian*, August 3, 2015, sec. World news. https://www.theguardian.com/world/2015/aug/03/europeans-who-welcome-migrants.

Bell, David M. *Rethinking Utopia: Place, Power, Affect*. New York/Abingdon: Routledge, 2017.

Belmont, Cynthia. "Ecosexuals in Appalachia: Identity, Community, and Counter-Discourse in *Goodbye Gauley Mountain*." *ISLE: Interdisciplinary Studies in Literature and Environment* 25, no. 4 (2018): 742–66.

Berner, Anna-Sofia. "Red Hook: The Hip New York Enclave Caught Between Gentrification and Climate Change." *The Guardian*, September 25, 2018, sec. Environment. https://www.theguardian.com/environment/2018/sep/25/red-hook-climate-change-floodplain-hurricane-sandy-gentrification.

Berners-Lee, Mike, and Duncan Clark. *The Burning Question: We Can't Burn Half the World's Oil, Coal, and Gas. So How Do We Quit?* Vancouver: Greystone Books, 2013.

Bernes, Jasper. "The Belly of the Revolution: Agriculture, Energy, and the Future of Communism." In *Materialism and the Critique of Energy*, edited by Brent Bellamy and Jeff Diamanti, 331–75. Edmonton and Chicago: MCM Prime Press, 2018.

Bettini, Giovanni. "Climate Barbarians at the Gate? A Critique of Apocalyptic Narratives on 'Climate Refugees.'" *Geoforum* 45 (2013): 63–72.

Bhambra, Gurminder K. *Connected Sociologies*. London: Bloomsbury Publishing, 2014.

Bhattacharya, Tithi, ed. *Social Reproduction Theory: Remapping Class, Recentring Oppression*. London: Pluto Press, 2017.

Biehl, Janet, and Peter Staudenmaier. *Ecofascism: Lessons from the German Experience*. Edinburgh and San Francisco: AK Press, 1995.

Blades, Lincoln Anthony. "Donald Trump Is Being Accused of Endorsing Police Brutality." *Teen Vogue*, July 30, 2017. https://www.teenvogue.com/story/donald-trump-nypd-speech-police-brutality.

Bloch, Ernst. *The Principle of Hope, Volume 1*. Translated by Neville Plaice, Stephen Plaice, and Paul Knight. Cambridge, MA: MIT Press, 1995.

Bonefeld, Werner. "Antisemitism and the Power of Abstraction: From Political Economy to Critical Theory." In *Antisemitism and the Constitution of Sociology*, edited by Marcel Stoetzler, 314–32. Lincoln: University of Nebraska Press, 2014.

Bonilla, Yarimar, and Marisol LeBrón. *Aftershocks of Disaster: Puerto Rico Before and After the Storm*. Chicago: Haymarket Books, 2019.

Bookchin, Debbie. "Radical Municipalism: The Future We Deserve." *ROAR Magazine*, July 21, 2017. https://roarmag.org/magazine/debbie-bookchin-municipalism-rebel-cities/.

Bookchin, Murray. *Post-Scarcity Anarchism*. Oakland: AK Press, 2004.

_____. *The Ecology of Freedom: The Emergence and Dissolution of Hierarchy*. Palo Alto, CA: Cheshire Books, 1982.

Briggs, Laura. *Reproducing Empire: Race, Sex, Science, and U.S. Imperialism in Puerto Rico*. Berkeley: University of California Press, 2003.

Brown, Oli. "Migration and Climate Change." International Organization for Migration Research Series. Geneva: International Organization of Migration, 2008. https://www.iom.cz/files/Migration_and_Climate_Change_-_IOM_Migration_Research_Series_No_31.pdf.

Caffentzis, George. "The Work/Energy Crisis and the Apocalypse." In *Midnight Oil: Work, Energy, War, 1973–1992*, edited by Midnight Notes Collective, 215–72. Brooklyn, NY: Autonomedia, 1992.

Campbell, Anneke, and Thomas Linzey. *We the People: Stories from the Community Rights Movement in the United States*. Oakland: PM Press, 2016.

Canadian Press, The. "Drowned Syrian Migrant Boy's Father Says He Blames Canada for Tragedy," September 10, 2015. https://www.theglobeandmail.com/news/national/drowned-syrian-boys-father-says-he-blames-canada-for-tragedy/article26313666/.

Carbon Tracker. "Wasted Capital and Stranded Assets." Carbon Tracker Initiative, April 19, 2013. https://www.carbontracker.org/reports/unburnable-carbon-wasted-capital-and-stranded-assets/.

Carney, Megan A., Ricardo Gomez, Katharyne Mitchell, and Sara Vannini. "Sanctuary Planet: A Global Sanctuary Movement for the Time of Trump." *Society & Space* (blog), 2017. http://societyandspace.org/2017/05/16/sanctuary-planet-a-global-sanctuary-movement-for-the-time-of-trump/.

Chrisafis, Angelique. "Farmer Given Suspended €3,000 Fine for Helping Migrants Enter France." *The Guardian*, February 10, 2017, sec. World news. https://www.theguardian.com/world/2017/feb/10/cedric-herrou-farmer-given-suspended-3000-fine-for-helping-migrants-enter-france.

Ciccariello-Maher, George. *Decolonizing Dialectics*. Durham: Duke University Press, 2017.

Ciplet, David, Mizan R. Khan, and J. Timmons Roberts. *Power in a Warming World: The New Global Politics of Climate Change and the Remaking of Environmental Inequality*. Cambridge, MA: MIT Press, 2015.

Ciplet, David, J. Timmons Roberts, and Mizan Khan. "The Politics of International Climate Adaptation Funding: Justice and Divisions in the Greenhouse." *Global Environmental Politics* 13, no. 1 (2013): 49–68.

Cleaver, Harry. "The Uses of an Earthquake." *Midnight Notes* 9 (1988): 10–14.

Cleveland, David. *Balancing on a Planet: The Future of Food and Agriculture*. Berkeley and Los Angeles: University of California Press, 2014.

Clover, Joshua. "The Roundabout Riots." *Versobooks.com* (blog), December 9, 2018. https://www.versobooks.com/blogs/4161-the-roundabout-riots.

Consterdine, Erica. "UK to Remain a Hostile Environment for Immigration Under Nebulous New Post-Brexit Policy." *The Conversation*, December 20, 2018. http://theconversation.com/uk-to-remain-a-hostile-environment-for-immigration-under-nebulous-new-post-brexit-policy-109095.

Costanza, Robert, Rudolf de Groot, Paul Sutton, Sander Van der Ploeg, Sharolyn J. Anderson, Ida Kubiszewski, Stephen Farber, and R. Kerry Turner. "Changes in the Global Value of Ecosystem Services." *Global Environmental Change* 26 (2014): 152–58.

Coulthard, Glen Sean. "Place Against Empire: Understanding Indigenous Anti-Colonialism." *Affinities: A Journal of Radical Theory, Culture, and Action* 4, no. 2 (2010): 79–83.

_____. *Red Skin, White Masks: Rejecting the Colonial Politics of Recognition*. Minneapolis: University of Minnesota Press, 2014.

crow, scott. *Black Flags and Windmills: Hope, Anarchy, and the Common Ground Collective*. Oakland: PM Press, 2011.

Cuboniks, Laboria. *The Xenofeminist Manifesto: A Politics for Alienation*. London and New York: Verso Books, 2018.

Dauvé, Gilles. "A Contribution to the Critique of Political Autonomy." *Libcom.org* (blog), October 26, 2008. http://libcom.org/library/a-contribution-critique-political-autonomy-gilles-dauve-2008.

Davis, Mike. *Late Victorian Holocausts: El Niño Famines and the Making of the Third World*. Reprint edition. London and New York: Verso Books, 2017.

Dawson, Ashley. *Extreme Cities: The Peril and Promise of Urban Life in the Age of Climate Change*. London and New York: Verso Books, 2017.

Day, Iyko. *Alien Capital: Asian Racialization and the Logic of Settler Colonial Capitalism*. Durham: Duke University Press, 2016.

Dean, Jodi. *The Communist Horizon*. London and New York: Verso Books, 2012.

Dehghan, Saeed Kamali. "Migrant Sea Route to Italy Is World's Most Lethal." *The Guardian*, September 10, 2017, sec. World news. https://www.theguardian.com/world/2017/sep/11/migrant-death-toll-rises-after-clampdown-on-east-european-borders.

Deleuze, Gilles. *Cinema 2: The Time Image*. Translated by Hugh Tomlinson and Robert Galeta. Minneapolis: University of Minnesota Press, 1989.

Deleuze, Gilles, and Félix Guattari. *A Thousand Plateaus: Capitalism and Schizophrenia*. Translated by Brian Massumi. Minneapolis: University of Minnesota Press, 1987.

_____. *Anti-Oedipus: Capitalism and Schizophrenia*. Translated by Robert Hurley, Mark Seem, and Helen Lane. Minneapolis: University of Minnesota Press, 1983.

Democracy Now! Staff. "Mass Graves of Immigrants Found in Texas, But State Says No Laws Were Broken." Democracy Now!, July 16, 2015. http://www.democracynow.org/2015/7/16/mass_graves_of_immigrants_found_in.

Dermansky, Julie. "Critics Say Louisiana 'Highjacked' Climate Resettlement Plan for Isle de Jean Charles Tribe." *DeSmogBlog* (blog), April 20, 2019. https://www.desmogblog.com/2019/04/20/critics-louisiana-highjacked-climate-resettlement-plan-isle-de-jean-charles-tribe.

Devereaux, Ryan. "Humanitarian Volunteer Scott Warren Reflects on the Borderlands and Two Years of Government Persecution." *The Intercept* (blog), November 23, 2019. https://theintercept.com/2019/11/23/scott-warren-verdict-immigration-border/.

Docs Not Cops. "#NHS70 – No Borders in Healthcare." *Docs Not Cops* (blog), July 5, 2018. http://www.docsnotcops.co.uk/nhs70-no-borders-in-healthcare/.

Doherty, Ben. "A Short History of Nauru, Australia's Dumping Ground for Refugees." *The Guardian*, August 9, 2016, sec. World news. https://www.theguardian.com/world/2016/aug/10/a-short-history-of-nauru-australias-dumping-ground-for-refugees.

Dorling, Danny. *Population 10 Billion*. London: Constable, 2013.

Dunbar-Ortiz, Roxanne. *An Indigenous Peoples' History of the United States*. Boston: Beacon Press, 2014.

Dyer-Witheford, Nick. *Cyber-Proletariat: Global Labour in the Digital Vortex*. London: Pluto Press, 2015.

Dyson, Tim. *Population and Development: The Demographic Transition*. London: Zed Books, 2013.

Edelman, Lee. *No Future: Queer Theory and the Death Drive*. Durham: Duke University Press, 2004.

EJOLT. "Phosphate Mining on Nauru | EJAtlas." Environmental Justice Atlas. Accessed April 25, 2019. https://ejatlas.org/conflict/phosphate-mining-on-nauru.

Endnotes Collective. *Endnotes 4: Unity in Separation*. London: Endnotes, 2015.

Estes, Nick. *Our History Is the Future: Standing Rock Versus the Dakota Access Pipeline, and the Long Tradition of Indigenous Resistance*. London and New York: Verso Books, 2019.

Fanon, Frantz. *The Wretched of the Earth*. Translated by Richard Philcox. New York: Grove Press, 2004.

Federici, Silvia. "Preoccupying." *The Occupied Times* (blog), October 25, 2014. https://theoccupiedtimes.org/?p=13482.

_____. *Re-Enchanting the World: Feminism and the Politics of the Commons*. Oakland: PM Press, 2018.

_____. *Revolution at Point Zero: Housework, Reproduction, and Feminist Struggle*. Oakland and Brooklyn: PM Press/Common Notions/Autonomedia, 2012.

Fisher, Mark. *Capitalist Realism: Is There No Alternative?* Winchester: Zero Books, 2009.

Food and Agriculture Organization of the United Nations. "FAO's Strategic Objective 1: Help Eliminate Hunger, Food Insecurity and Malnutrition," 2015. http://www.fao.org/3/a-au829e.pdf.

Forchtner, Bernhard, ed. *The Far Right and the Environment: Politics, Discourse and Communication*. Abingdon and New York: Routledge, 2019.

Foster, John Bellamy. *Ecology Against Capitalism*. New York: Monthly Review Press, 2002.

Foster, John Bellamy, Richard York, and Brett Clark. *The Ecological Rift: Capitalism's War on the Earth*. New York: Monthly Review Press, 2011.

Frank, Jason. *Constituent Moments: Enacting the People in Postrevolutionary America*. Durham: Duke University Press, 2009.

Freedman, Andrew. "Climate Scientists Refute 12-Year Deadline to Curb Global Warming." *Axios*, January 22, 2019. https://www.axios.com/climate-change-scientists-comment-ocasio-cortez-12-year-deadline-c4ba1f99-bc76-42ac-8b93-e4eaa926938d.html.

Friedman, Thomas L. "Trump Is Wasting Our Immigration Crisis." *The New York Times*, April 25, 2019, sec. Opinion. https://www.nytimes.com/2019/04/23/opinion/trump-immigration-border-wall.html.

Fritz, Charles E. "Disasters and Mental Health: Therapeutic Principles Drawn From Disaster Studies." Historical and Comparative Disaster Series. University of Delaware Disaster Research Center, 1996. http://udspace.udel.edu/handle/19716/1325.

Gane, Nicholas. "When We Have Never Been Human, What Is to Be Done? Interview with Donna Haraway." *Theory, Culture & Society* 23, no. 7–8 (2006): 135–58.

Gassmann, Aaron J., Jennifer L. Petzold-Maxwell, Eric H. Clifton, Mike W. Dunbar, Amanda M. Hoffmann, David A. Ingber, and Ryan S. Keweshan. "Field-Evolved Resistance by Western Corn Rootworm to Multiple Bacillus Thuringiensis Toxins in Transgenic Maize." *Proceedings of the National Academy of Sciences* 111, no. 14 (2014): 5141–46.

Gilmore, Inigo. "The Myth of Haiti's Lawless Streets." *The Guardian*, January 20, 2010, sec. Opinion. https://www.theguardian.com/commentisfree/2010/jan/20/haiti-aid-agency-security.

Gilmore, Ruth Wilson. "Abolition Geography and the Problem of Innocence." In *Futures of Black Radicalism*, edited by Gaye Theresa Johnson and Alex Lubin, 225–40. London and New York: Verso Books, 2017.

_____. *Golden Gulag: Prisons, Surplus, Crisis, and Opposition in Globalizing California*. Berkeley: University of California Press, 2007.

Gilpin, Emilee. "Urgency in Climate Change Advocacy Is Backfiring, Says Citizen Potawatomi Nation Scientist." *National Observer*, February 15, 2019. https://www.nationalobserver.com/2019/02/15/features/urgency-climate-change-advocacy-backfiring-says-citizen-potawatomi-nation.

Goldenberg, Suzanne. "Pentagon: Global Warming Will Change How US Military Trains and Goes to War." *The Guardian*, October 13, 2014, sec. Environment. https://www.theguardian.com/environment/2014/oct/13/pentagon-global-warming-will-change-how-us-military-trains-and-goes-to-war.

Gordon, Uri. "Anarchism and Nationalism." In *Brill's Companion to Anarchism and Philosophy*, 196–215. Leiden: BRILL, 2018.

Gossett, Che. "Blackness, Animality, and the Unsovereign." *Verso Books* (blog), September 8, 2015. https://www.versobooks.com/blogs/2228-che-gossett-blackness-animality-and-the-unsovereign.

Grandin, Greg. *The End of the Myth: From the Frontier to the Border Wall in the Mind of America*. New York: Metropolitan Books, 2019.

Griffin, Roger. *The Nature of Fascism*. London: Routledge, 1993.

Guiot, Joel, and Wolfgang Cramer, "Climate change: The 2015 Paris Agreement Thresholds and Mediterranean Basin Ecosystems," *Science*, 354 (2016): 465-468.

Gustin, Georgina. "Florida's Migrant Farm Workers Struggle After Hurricane Damaged Homes, Crops." *InsideClimate News*, October 17, 2017. https://insideclimatenews.org/news/16102017/hurricanes-florida-agriculture-migrant-farm-workers-jobs-crop-loss.

Haghamed, Naser. "The Muslim World Has to Take Climate Action." *Al-Jazeera*, November 4, 2016. https://www.aljazeera.com/indepth/opinion/2016/11/muslim-world-climate-action-161103101248390.html.

Haidt, Jonathan. "When and Why Nationalism Beats Globalism." *Policy: A Journal of Public Policy and Ideas* 32, no. 3 (2016): 46–53.

Haraway, Donna. "The Promises of Monsters: A Regenerative Politics for Inappropriate/d Others." In *Cultural Studies*, edited by Lawrence Grossberg, Cary Nelson, and Paula A. Treichler, 295–337. New York: Routledge, 1992.

Haraway, Donna. "Cyborgs for Earthly Survival!" In *London Review of Books* 39, No. 13, June 29, 2017, sec. Letters. https://www.lrb.co.uk/v39/n13/letters.

———. *Manifestly Haraway*. Minneapolis: University of Minnesota Press, 2016.

———. *Staying with the Trouble: Making Kin in the Chthulucene*. Durham: Duke University Press, 2016.

———. "Tentacular Thinking: Anthropocene, Capitalocene, Chthulucene." *e-flux journal* 75 (September 2016). https://www.e-flux.com/journal/75/67125/tentacular-thinking-anthropocene-capitalocene-chthulucene/.

———. *When Species Meet*. Minneapolis: University of Minnesota Press, 2008.

Haraway, Donna, and Martha Kenny. "Anthropocene, Capitalocene, Chthulhucene: Donna Haraway in Conversation with Martha Kenney." In *Art in the Anthropocene: Encounters among Aesthetics, Politics, Environments and Epistemologies*, edited by Heather Davis and Etienne Turpin, 255–69. London: Open Humanities Press, 2015.

Hardin, Garrett. "Extensions of 'the Tragedy of the Commons.'" *Science* 280, no. 5364 (1998): 682–83.

———. "Lifeboat Ethics: The Case Against Helping the Poor." *Psychology Today* 8 (1974): 38–43.

———. "The Tragedy of the Commons." *Science* 162, no. 3859 (1968): 1243–48.

Harney, Stefano, and Fred Moten. *The Undercommons: Fugitive Planning & Black Study*. Wivenhoe, New York, Port Watson: Minor Compositions, 2013.

Hartmann, Betsy. *Reproductive Rights and Wrongs*. Boston: South End Press, 1999.

Harvey, David, and Donna Haraway. "Nature, Politics, and Possibilities: A Debate and Discussion with David Harvey and Donna Haraway." *Environment and Planning D: Society and Space* 13, no. 5 (1995): 507–27.

Hatherley, Owen. "Lash Out and Cover Up." *Radical Philosophy* 157 (2009): 2–7.

Heglar, Mary Annaïse. "Climate Change Ain't the First Existential Threat." *Medium* (blog), February 18, 2019. https://medium.com/s/story/sorry-yall-but-climate-change-ain-t-the-first-existential-threat-b3c999267aa0.

———. "When Climate Change Broke My Heart and Forced Me to Grow Up." *Medium* (blog), October 10, 2018. https://medium.com/@maryheglar/when-climate-change-broke-my-heart-and-forced-me-to-grow-up-dcffc8d763b8.

Heinrich, Michael. "Crisis Theory, the Law of the Tendency of the Profit Rate to Fall, and Marx's Studies in the 1870s." *Monthly Review* 64, no. 11 (2013): 15–31.

Hernández, César Cuauhtémoc García. "This Man Could Go to Jail for 20 Years for Giving Migrants Food and Water." *The Guardian*, May 10, 2019, sec. Opinion. https://www.theguardian.com/commentisfree/2019/may/10/scott-daniel-warren-migrants-food-water.

Hill, David. "Ecuador Pursued China Oil Deal While Pledging to Protect Yasuni, Papers Show." *The Guardian*, February 19, 2014, sec. Environment. https://www.theguardian.com/environment/2014/feb/19/ecuador-oil-china-yasuni.

Hintze, Thomas. "Homeland Security Study Praises Occupy Sandy, With Murky Intentions." *Truthout*, April 2, 2014. https://truthout.org/articles/dhs-study-praises-occupy-sandy-with-murky-intentions/.

Holloway, John. "The Concept of Power and the Zapatistas." *Common Sense* 19 (1996): 20–27.

Hood, Stephen l'Argent. "Autochthony, Promised Land, and Exile: Athens and Jerusalem Revisited." PhD Thesis, 2006. https://scholarship.rice.edu/handle/1911/18918.

Imarisha, Walidah, and adrienne maree brown, eds. *Octavia's Brood: Science Fiction Stories from Social Justice Movements*. Oakland, CA: AK Press, 2015.

Intergovernmental Panel on Climate Change. "Global Warming of 1.5 °C —." UNEP, 2018. https://www.ipcc.ch/sr15/.

———. "Impacts, Adaption, and Vulnerability, Part A: Global and Sectoral Aspects. Contribution of Working Group II to the Fifth Assessment Report of the Intergovernmental Panel on Climate Change." New York: Cambridge University Press, December 29, 2014.

Jahn, Beate. *The Cultural Construction of International Relations: The Invention of the State of Nature*. Hampshire and New York: Palgrave Macmillan, 2000.

James, C.L.R. "Every Cook Can Govern." *Correspondence*, June 1956. https://www.marxists.org/archive/james-clr/works/1956/06/every-cook.htm.

Japanese and Korean Exclusion League. "Asiatic Coolie Invasion." *Organized Labor*, 1906. http://www.sfmuseum.net/1906.2/invasion.html.

Jared. "Amongst the Rubble: A Look at the Christchurch Earthquake from the Bottom Up." *Libcom.org* (blog), September 16, 2010. http://libcom.org/news/amongst-rubble-look-christchurch-earthquake-bottom-16092010.

Jones, Graham. *The Shock Doctrine of the Left*. Hoboken, NJ: John Wiley & Sons, 2018.

Jones, Reece. *Violent Borders: Refugees and the Right to Move*. London and New York: Verso Books, 2016.

Kane, Katie. "Nits Make Lice: Drogheda, Sand Creek, and the Poetics of Colonial Extermination." *Cultural Critique*, no. 42 (1999): 81–103.

Kanngieser, Anja. "Climate Change: Nauru's Life on the Frontlines." *The Conversation*, October 21, 2018. http://theconversation.com/climate-change-naurus-life-on-the-frontlines-105219.

Kelbert, Alexandra Wanjiku. "Climate Change Is a Racist Crisis: That's Why Black Lives Matter Closed an Airport." *The Guardian*, September 6, 2016, sec. Opinion. https://www.theguardian.com/commentisfree/2016/sep/06/climate-change-racist-crisis-london-city-airport-black-lives-matter.

Kermoal, Nathalie, and Isabel Altamirano-Jiménez. *Living on the Land: Indigenous Women's Understanding of Place*. Edmonton: Athabasca University Press, 2016.

King, Natasha. *No Borders: The Politics of Immigration Control and Resistance*. London: Zed Books Ltd., 2016.

Kingsley, Patrick. "Passport, Lifejacket, Lemons: What Syrian Refugees Pack for the Crossing to Europe." *The Guardian*, September 4, 2015. http://www.theguardian.com/world/ng-interactive/2015/sep/04/syrian-refugees-pack-for-the-crossing-to-europe-crisis.

Kingsnorth, Paul. "The Lie of the Land: Does Environmentalism Have a Future in the Age of Trump?," *The Guardian*, March 18, 2017, sec. Books. http://www.theguardian.com/books/2017/mar/18/the-new-lie-of-the-land-what-future-for-environmentalism-in-the-age-of-trump.

Kitsantonis, Niki. "Anarchists Fill Services Void Left by Faltering Greek Governance." *The New York Times*, May 22, 2017, sec. World. https://www.nytimes.com/2017/05/22/world/europe/greece-athens-anarchy-austerity.html.

Klein, Naomi. *The Battle For Paradise: Puerto Rico Takes on the Disaster Capitalists*. Chicago: Haymarket Books, 2018.

_____. *The Shock Doctrine: The Rise of Disaster Capitalism*. New York: Picador, 2007.

_____. *This Changes Everything: Capitalism vs. The Climate*. New York: Simon & Schuster, 2014.

Kneale, James. "From Beyond: HP Lovecraft and the Place of Horror." *Cultural Geographies* 13, no. 1 (2006): 106–26.

Kundnani, Arun. "Blind Spot? Security Narratives and Far-Right Violence in Europe." ICCT Research Paper. The Hague: The International Centre for Counter-Terrorism (ICCT), 2012. https://www.icct.nl/download/file/ICCT-Kundnani-Blind-Spot-June-2012.pdf.

Kunzru, Hari. "You Are Cyborg." *Wired Magazine* 5, no. 2 (1997): 1–7.

La Vía Campesina. "Declaration of Rights of Peasants — Women and Men," 2009. https://viacampesina.net/downloads/PDF/EN-3.pdf.

La Vía Campesina, European Coordination. "Food Sovereignty Now! A Guide to Food Sovereignty." Brussels: European Coordination Vía Campesina, 2018. https://viacampesina.org/en/wp-content/uploads/sites/2/2018/02/Food-Sovereignty-A-guide-Low-Res-Vresion.pdf.

Laclau, Ernesto. *On Populist Reason*. London and New York: Verso Books, 2005.

Leahy, Stephen. "Climate Change Impacts Worse than Expected, Global Report Warns." *National Geographic*, October 7, 2018. https://www.nationalgeographic.com/environment/2018/10/ipcc-report-climate-change-impacts-forests-emissions/.

LeBrón, Marisol, and Javier Arbona. "Post-Maria Puerto Rico." *The Funambulist Magazine*, April 2018. https://thefunambulist.net/articles/guest-columnists. [Subscription required.]

Lees, Loretta. "The Urban Injustices of New Labor's 'New Urban Renewal': The Case of the Aylesbury Estate in London." *Antipode* 46, no. 4 (2014): 921–47.

Leverink, Joris. "Murray Bookchin and the Kurdish Resistance." *ROAR Magazine*, August 9, 2015. https://roarmag.org/essays/bookchin-kurdish-struggle-ocalan-rojava/.

Levi, Primo. *If This Is a Man and The Truce*. London: Abacus, 2003.

Levin, Kelly. "8 Things You Need to Know About the IPCC 1.5°C Report." *World Resources Institute*, October 7, 2018. https://www.wri.org/blog/2018/10/8-things-you-need-know-about-ipcc-15-c-report.

Lewis, Sophie. *Full Surrogacy Now: Feminism Against Family*. London and New York: Verso Books, 2019.

Liberti, Stefano. *Land Grabbing: Journeys in the New Colonialism*. Translated by Enda Flannelly. London and New York: Verso Books, 2013.

Lindeman, Raymond L. "The Trophic-Dynamic Aspect of Ecology." *Ecology* 23, no. 4 (1942): 399–417.

Lotter, Donald W. "Organic Agriculture." *Journal of Sustainable Agriculture* 21, no. 4 (2003): 59–128.

Lowndes, Joe, and Dorian Warren. "Occupy Wall Street: A Twenty-First Century Populist Movement?" *Dissent Magazine*, October 21, 2011. https://www.dissentmagazine.org/online_articles/occupy-wall-street-a-twenty-first-century-populist-movement.

Lynas, Mark. *Six Degrees: Our Future on a Hotter Planet*. Washington, DC: National Geographic Books, 2008.

Mackay, James, and David Stirrup. "There Is No Such Thing as an 'Indigenous' Briton." *The Guardian*, December 20, 2010, sec. Opinion. https://www.theguardian.com/commentisfree/2010/dec/20/indigenous-britons-far-right.

Magdoff, Fred, and John Bellamy Foster. *What Every Environmentalist Needs to Know About Capitalism: A Citizen's Guide to Capitalism and the Environment*. New York: Monthly Review Press, 2011.

Malm, Andreas. *The Progress of This Storm: Nature and Society in a Warming World*. London and New York: Verso Books, 2018.

Mann, Geoff, and Joel Wainwright. *Climate Leviathan: A Political Theory of Our Planetary Future*. London and New York: Verso Books, 2018.

Marx, Karl. *Capital: A Critique of Political Economy, Volume I*. Translated by Ben Fowkes. London: Penguin Classics, 1976.

_____. *Capital: A Critique of Political Economy, Volume III*. Translated by David Fernbach. London: Penguin Classics, 1981.

_____. "Economic Manuscripts: Marx's Economic Manuscripts of 1861–63." Accessed April 26, 2019. https://marxists.catbull.com/archive/marx/works/1861/economic/ch37.htm.

_____. *Grundrisse*. London: Penguin, 1973.

_____. "Critique of the Gotha Program." In *The Political Writings*, 1023–43. London and New York: Verso Books, 2019.

Marx, Karl, and Friedrich Engels. *German Ideology, Part 1 and Selections from Parts 2 and 3*. Edited by Christopher John Arthur. New York: International Publishers, 1970.

Max Planck Society. "Climate-Exodus Expected in the Middle East and North Africa." Phys.org, May 2, 2016. https://phys.org/news/2016-05-climate-exodus-middle-east-north-africa.html.

Mbembe, Achille. "Necropolitics." Translated by Libby Meintjes. *Public Culture* 15, no. 1 (2003): 11–40.

McGreal, Chris, and Esther Addley. "Haiti Aid Agencies Warn: Chaotic and Confusing Relief Effort Is Costing Lives." *The Guardian*, January 18, 2010, sec. World news. https://www.theguardian.com/world/2010/jan/18/haiti-aid-distribution-confusion-warning.

McKibben, Bill. "Global Warming's Terrifying New Math." *Rolling Stone*, July 19, 2012. https://www.rollingstone.com/politics/politics-news/global-warmings-terrifying-new-math-188550/.

_____. *Oil and Honey: The Education of an Unlikely Activist*. New York: Times Books, 2013.

Medel, China. "Abolitionist Care in the Militarized Borderlands." *South Atlantic Quarterly* 116, no. 4 (October 1, 2017): 873–83.

Michelle, Josie. "Against the New Vitalism." *New Socialist* (blog), March 10, 2019. https://newsocialist.org.uk/against-the-new-vitalism/.

Midnight Notes Collective. "The New Enclosures." In *Midnight Oil: Work, Energy, War, 1973–1992*, edited by Midnight Notes Collective, 317–33. Brooklyn, NY: Autonomedia, 1992.

Mies, Maria. *Patriarchy and Accumulation on a World Scale: Women in the International Division of Labour*. 2nd Edition. London: Zed Books, 1998.

Mignolo, Walter. *The Darker Side of Western Modernity: Global Futures, Decolonial Options*. Durham and London: Duke University Press, 2011.

Miller, Todd. *Storming the Wall: Climate Change, Migration, and Homeland Security*. San Francisco: City Lights Books, 2017.

Minca, Claudio, and Alexandra Rijke. "Walls! Walls! Walls!" *Society & Space* (blog), April 18, 2017. http://societyandspace.org/2017/04/18/walls-walls-walls/.

Mitropoulos, Angela. *Contract and Contagion: From Biopolitics to Oikonomia*. Wivenhoe, New York, Port Watson: Minor Compositions, 2012.

_____. "Corbynomics, Moral Economy and Saving Capitalism." *S0metim3s* (blog), July 17, 2018. https://s0metim3s.com/2018/07/17/corbynomics/.

_____. "Lifeboat Capitalism, Catastrophism, Borders." *Dispatches Journal*, November 19, 2018. http://dispatchesjournal.org/articles/162/.

_____. "On Borders / Race / Fascism / Labour / Precarity / Feminism / Etc." *BASE* (blog), October 29, 2016. https://www.basepublication.org/?p=107.

Møhring Reestorff, Camilla. "LEGO: Everything Is NOT Awesome!" *Conjunctions: Transdisciplinary Journal of Cultural Participation*. 2, no. 1 (2015): 21–43.

Moore, Jason W. *Capitalism in the Web of Life: Ecology and the Accumulation of Capital*. London and New York: Verso Books, 2015.

_____. "Ecology, Capital, and the Nature of Our Times: Accumulation and Crisis in the Capitalist World-Ecology." *Journal of World-Systems Research* 17, no. 1 (2011): 108–47.

Moorti, Sujata. "A Queer Romance with the Hijra." *QED: A Journal in GLBTQ Worldmaking*, 3, no. 2 (2016): 18–34.

Morris, Aldon D. *The Origins of the Civil Rights Movement: Black Communities Organizing for Change*. New York: Free Press, 1986.

Moten, Fred. "Blackness and Poetry." *ARCADE*, no. 55 (2015). https://arcade.stanford.edu/content/blackness-and-poetry-0.

Mozur, Paul. "China Scrutinizes 2 Apple Suppliers in Pollution Probe." *Wall Street Journal*, August 4, 2013, sec. Business. https://www.wsj.com/articles/SB10001424127887323420604578648002283373528.

Murphy, Michelle. *The Economization of Life*. Durham: Duke University Press, 2017.

Myers, Norman. "Environmental Refugees: An Emergent Security Issue." Prague: Organization for Security and Co-operation in Europe, May 25, 2005. https://www.osce.org/eea/14851.

Natural Resources Defense Council. "Threats on Tap: Drinking Water Violations in Puerto Rico." NRDC, May 2, 2017. https://www.nrdc.org/resources/threats-tap-drinking-water-violations-puerto-rico.

Nature Climate Change. "Carbon Bubble Toil and Trouble." *Nature Climate Change* 4, no. 4 (April 2014): 229.

Nature eds. "How to Feed a Hungry World." *Nature* 466, no. 7306 (2010): 531–32.

Newell, Peter, and Matthew Paterson. *Climate Capitalism: Global Warming and the Transformation of the Global Economy*. Cambridge: Cambridge University Press, 2010.

Neyrat, Frédéric. *The Unconstructable Earth: An Ecology of Separation*. New York: Fordham University Press, 2018.

Nixon, Rob. "Naomi Klein's 'This Changes Everything.'" *The New York Times*, November 6, 2014, sec. Books. https://www.nytimes.com/2014/11/09/books/review/naomi-klein-this-changes-everything-review.html.

_____. *Slow Violence and the Environmentalism of the Poor*. Cambridge, MA: Harvard University Press, 2011.

"Occupy Sandy Staten Island Tool Library | Occupy Sandy Recovery." Accessed April 26, 2019. http://occupysandy.net/?projects=occupy-sandy-staten-island-tool-library.

O'Connor, James. "Capitalism, Nature, Socialism: A Theoretical Introduction." *Capitalism Nature Socialism* 1, no. 1 (January 1, 1988): 11–38.

Ollman, Bertell. *Dance of the Dialectic: Steps in Marx's Method*. Champaign: University of Illinois Press, 2003.

Oreskes, Naomi, and Erik M. Conway. *Merchants of Doubt: How a Handful of Scientists Obscured the Truth on Issues from Tobacco Smoke to Global Warming*. New York: Bloomsbury Publishing USA, 2011.

Ostler, Jeffrey. *The Lakotas and the Black Hills: The Struggle for Sacred Ground*. New York: Penguin Books, 2011.

Ostrom, Elinor. *Governing the Commons: The Evolution of Institutions for Collective Action*. Cambridge: Cambridge University Press, 1990.

Out of the Woods. "Class Struggles, Climate Change, and the Origins of Modern Agriculture." *Libcom.org* (blog), August 18, 2014. https://libcom.org/blog/class-struggles-climate-change-origins-modern-agriculture-18082014.

_____. "Climate, Class, and the Neolithic Revolution." *Libcom.org* (blog), June 9, 2014. https://libcom.org/blog/climate-class-neolithic-revolution-09062014.

_____. "Disaster Communism Part 1 – Disaster Communities." *Libcom.org* (blog), May 8, 2014. http://libcom.org/blog/disaster-communism-part-1-disaster-communities-08052014.

_____. "Disaster Communism Part 2 – Communisation and Concrete Utopia." *Libcom.org* (blog), May 14, 2014. http://libcom.org/blog/disaster-communism-part-2-communisation-concrete-utopia-14052014.

_____. "Disaster Communism Part 3 – Logistics, Repurposing, Bricolage," *Libcom.org* (blog), May 22, 2014. http://libcom.org/blog/disaster-communism-part-3-logistics-repurposing-bricolage-22052014.

_____. "Extinction Rebellion: Not the Struggle We Need, Pt. 1." *Libcom.org* (blog), July 19, 2019. http://libcom.org/blog/extinction-rebellion-not-struggle-we-need-pt-1-19072019.

_____. "Extinction Rebellion: Not the Struggle We Need, Pt. 2." *Libcom.org* (blog), October 31, 2019. http://libcom.org/blog/xr-pt-2-31102019.

_____. "No Future." *The Occupied Times* (blog), October 22, 2014. https://theoccupiedtimes.org/?p=13483.

Patel, Raj, and Jason W. Moore. *A History of the World in Seven Cheap Things: A Guide to Capitalism, Nature, and the Future of the Planet*. Oakland: University of California Press, 2017.

Pellow, David N., and Lisa Sun-Hee Park. *The Silicon Valley of Dreams: Environmental Injustice, Immigrant Workers, and the High-Tech Global Economy*. New York: NYU Press, 2002.

Perelman, Michael. *The Invention of Capitalism: Classical Political Economy and the Secret History of Primitive Accumulation*. Durham: Duke University Press, 2000.

Piggot, Georgia, Peter Erickson, Michael Lazarus, and Harro van Asselt. *Addressing Fossil Fuel Production Under the UNFCCC: Paris and Beyond*. Seattle, WA: Stockholm Environment Institute, 2017.

Plumer, Brad. "Ecuador Asked the World to Pay It Not to Drill for Oil. The World Said No." *Washington Post*, August 16, 2013. https://www.washingtonpost.com/news/wonk/wp/2013/08/16/ecuador-asked-the-world-to-pay-it-not-to-drill-for-oil-the-world-said-no/.

Plumwood, Val. *Feminism and the Mastery of Nature*. London: Routledge, 1993.

Polanyi, Karl. *The Great Transformation: The Political and Economic Origins of Our Time*. Boston: Beacon Press, 2001.

Refugee Action Coalition. "Protests Escalate on Nauru," April 6, 2016. http://www.refugeeaction.org.au/?p=4859.

Puar, Jasbir K. "'I Would Rather Be a Cyborg than a Goddess': Becoming-Intersectional in Assemblage Theory." *PhiloSOPHIA* 2, no. 1 (2012): 49–66.

_____. *Terrorist Assemblages: Homonationalism in Queer Times*. Durham: Duke University Press, 2007.

_____. *The Right to Maim: Debility, Capacity, Disability*. Durham: Duke University Press, 2017.

Pulido, Laura. "Geographies of Race and Ethnicity II: Environmental Racism, Racial Capitalism and State-Sanctioned Violence." *Progress in Human Geography* 41, no. 4 (2017): 524–33.

Ramnath, Maia. *Decolonizing Anarchism: An Antiauthoritarian History of India's Liberation Struggle*. Oakland: AK Press, 2012.

"'Record Hate Crimes' after EU Referendum." *BBC News*, February 15, 2017, sec. UK. https://www.bbc.com/news/uk-38976087.

Reicher, Steve, and Alex Haslam. "A 'Migrant' Is Not a Migrant by Any Other Name." *The Psychologist*, September 11, 2015. https://thepsychologist.bps.org.uk/migrant-not-migrant-any-other-name.

Revolts Now. "Disaster Communism & Anarchy in the Streets." *Revolts Now* (blog), April 10, 2011. https://revoltsnow.wordpress.com/2011/04/10/166/.

Ricchi. "Dreaming With Our Hands: On Autonomy, In(Ter)Dependence, and the Regaining of the Commons." *Mutual Aid Disaster Relief* (blog), January 9, 2018. https://mutualaiddisasterrelief.org/dreaming-with-our-hands-on-autonomy-interdependence-and-the-regaining-of-the-commons/.

Riofrancos, Thea. "Democracy Without the People." *n+1* (blog), February 6, 2017. https://nplusonemag.com/online-only/online-only/democracy-without-the-people/.

Rising Tide North America. "Growing the Roots to Weather the Storm: Reflections on the Global Movement for Climate Justice." *Growing the Roots to Weather the Storm*, 2014. http://growingdeeproots.risingtidenorthamerica.org/?cat=6.

Roberts, Dorothy E. *Killing the Black Body: Race, Reproduction, and the Meaning of Liberty*. New York: Vintage Books, 1999.

Rosen, Michael. "Fascism: I Sometimes Fear. . . ." *Michael Rosen* (blog), May 18, 2014. http://michaelrosenblog.blogspot.com/2014/05/fascism-i-sometimes-fear.html.

Rutherford, Adam. *Creation: How Science Is Reinventing Life Itself*. New York: Current, 2013.

Salmón, Enrique. "Kincentric Ecology: Indigenous Perceptions of the Human-Nature Relationship." *Ecological Applications* 10, no. 5 (2000): 1327–32.

International Women's Strike USA. "San Francisco – Gender Strike! Bay Area," March 8, 2017. https://www.womenstrikeus.org/event/san-francisco-gender-strike-bay-area/.

Schlembach, Raphael. *Against Old Europe: Critical Theory and Alter-Globalization Movements.* Abingdon and New York: Routledge, 2016.

schoolsabc. "We Won! DfE Are Ending the Nationality School Census!" *Against Borders for Children* (blog), April 10, 2018. https://www.schoolsabc.net/2018/04/we-won/.

Schumacher, E. F. *Small Is Beautiful: Economics as If People Mattered.* New York: Harper, 1973.

Scofield, Be. "How Derrick Jensen's Deep Green Resistance Supports Transphobia." *Decolonizing Yoga* (blog), May 24, 2013. http://www.decolonizingyoga.com/how-derrick-jensens-deep-green-resistance-supports-transphobia/.

Scott, James C. *Seeing Like a State: How Certain Schemes to Improve the Human Condition Have Failed.* New Haven: Yale University Press, 1998.

Selby, Jan, Omar S. Dahi, Christiane Fröhlich, and Mike Hulme. "Climate Change and the Syrian Civil War Revisited." *Political Geography* 60 (2017): 232–44.

Sen, Amartya. *Poverty and Famines: An Essay on Entitlement and Deprivation.* Oxford: Oxford University Press, 1982.

Serres, Michel. *Malfeasance: Appropriation Through Pollution?* Translated by Anne-Marie Feenberg-Dibon. Stanford: Stanford University Press, 2011.

Seufert, Verena, Navin Ramankutty, and Jonathan A. Foley. "Comparing the Yields of Organic and Conventional Agriculture." *Nature* 485, no. 7397 (2012): 229.

Shiva, Vandana. *Soil Not Oil: Environmental Justice in an Age of Climate Crisis.* Berkeley: North Atlantic Books, 2015.

———. "Why the Government Is Right in Controlling the Price of Monsanto's Bt Cotton Seeds." *Scroll.In*, August 22, 2016. https://scroll.in/article/814476/why-the-government-is-right-in-controlling-the-price-of-monsantos-bt-cotton-seeds.

Silva, Denise Ferreira da. "Unpayable Debt: Reading Scenes of Value against the Arrow of Time." In *The Documenta 14 Reader*, edited by Quinn Latimer and Adam Szymczyk, 81–113. München, London, New York: Prestel, 2017.

Simpson, Leanne Betasamosake. *As We Have Always Done: Indigenous Freedom through Radical Resistance.* Minneapolis: University of Minnesota Press, 2017.

Sluijs, J. P. van der, V. Amaral-Rogers, L. P. Belzunces, M. F. I. J. Bijleveld van Lexmond, J-M. Bonmatin, M. Chagnon, C. A. Downs, et al. "Conclusions of the Worldwide Integrated Assessment on the Risks of Neonicotinoids and Fipronil to Biodiversity and Ecosystem Functioning." *Environmental Science and Pollution Research* 22, no. 1 (January 1, 2015): 148–54.

Smith, Helena. "Forgotten Inside Greece's Notorious Camp for Child Refugees." *The Guardian*, September 10, 2016, sec. World news. https://www.theguardian.com/world/2016/sep/10/child-refugees-greece-camps.

Smith, Neil. "There's No Such Thing as a Natural Disaster." Understanding Katrina: Perspectives from the Social Sciences, June 11, 2006. https://items.ssrc.org/understanding-katrina/theres-no-such-thing-as-a-natural-disaster/.

———. *Uneven Development: Nature, Capital, and the Production of Space.* Athens, GA: University of Georgia Press, 2008.

Smith-Spangler, Crystal, Margaret L. Brandeau, Grace E. Hunter, J. Clay Bavinger, Maren Pearson, Paul J. Eschbach, Vandana Sundaram, Hau Liu, Patricia Schirmer, and Christopher Stave. "Are Organic Foods Safer or Healthier than Conventional Alternatives?: A Systematic Review." *Annals of Internal Medicine* 157, no. 5 (2012): 348–66.

s.nappalos. "Responding to the Growing Importance of the State in the Workers' Movement." *Libcom.org* (blog), February 4, 2014. http://libcom.org/blog/responding-growing-importance-state-workers-movement-04022014.

Solidarity Federation. *Fighting for Ourselves: Anarcho-Syndicalism and the Class Struggle*, October 27, 2012. https://libcom.org/library/fighting-ourselves-anarcho-syndicalism-class-struggle-solidarity-federation.

Solnit, Rebecca. *A Paradise Built in Hell: The Extraordinary Communities That Arise in Disaster*. New York: Penguin, 2010.

spitzenprodukte. "Viva Miuccia! Cursory Notes on the Political T-Shirt." *Libcom.org* (blog), June 18, 2014. http://libcom.org/blog/viva-miuccia-cursory-notes-political-t-shirt-18062014.

Spratt, David, and Philip Sutton. *Climate Code Red: The Case for Emergency Action*. Carlton North, Vic: Scribe Publications, 2008.

Srnicek, Nick, and Alex Williams. *Inventing the Future: Postcapitalism and a World Without Work*. London and New York: Verso Books, 2015.

Stabroek News. "Fighting Complacency towards Climate Change." *Stabroek News* (blog), October 20, 2018. https://www.stabroeknews.com/2018/opinion/editorial/10/20/fighting-complacency-towards-climate-change/.

Staff, Guardian. "Letters: Many Greens Worried by High Immigration." *The Guardian*, July 25, 2013, sec. Politics. https://www.theguardian.com/politics/2013/jul/25/greens-worried-high-immigration.

Stothard, Michael. "Marine Le Pen Uses Environmental Issue to Broaden Appeal." *Financial Times*, January 26, 2017. https://www.ft.com/content/613eeb24-e3fc-11e6-9645-c9357a75844a.

Sua'ali'i, Tamasailau. "Samoans and Gender: Some Reflections on Male, Female and Fa'afafine Gender Identities." In *Tangata O Te Moana Nui: The Evolving Identities of Pacific Peoples in Aotearoa/New Zealand*, edited by Cluny Macpherson, Paul Spoonley, and Melani Anae, 160–80. Palmerston North: Dunmore Press, 2001.

TallBear, Kim. "Genomic Articulations of Indigeneity." *Social Studies of Science* 43, no. 4 (2013): 509–33.

Taylor, Dorceta E. *Toxic Communities: Environmental Racism, Industrial Pollution, and Residential Mobility*. New York: NYU Press, 2014.

"The Paris Agreement | UNFCCC." Accessed April 26, 2019. https://unfccc.int/process-and-meetings/the-paris-agreement/the-paris-agreement.

Thompson, Matt. "Staying with the Trouble: Making Kin in the Chthulucene [Review]." *Savage Minds* (blog), November 18, 2016. https://savageminds.org/2016/11/18/staying-with-the-trouble-making-kin-in-the-chthulucene-review/.

Travis, Alan. "Immigration Bill: Theresa May Defends Plans to Create 'Hostile Environment.'" *The Guardian*, October 10, 2013, sec. Politics. https://www.theguardian.com/politics/2013/oct/10/immigration-bill-theresa-may-hostile-environment.

Tronti, Mario. *Workers and Capital*. London and New York: Verso, 2019.

Turhan, Ethemcan, and Marco Armiero. "Cutting the Fence, Sabotaging the Border: Migration as a Revolutionary Practice." *Capitalism Nature Socialism* 28, no. 2 (2017): 1–9.

UNHCR. "Global Trends: Forced Displacement in 2017." Geneva: UNHCR, 2018. https://www.unhcr.org/5b27be547.pdf.

UNISDR. "Ten-Year Review Finds 87% of Disasters Climate-Related." Accessed April 25, 2019. https://www.unisdr.org/archive/42862.

United Nations High Commissioner for Refugees. "UNHCR: Total Number of Syrian Refugees Exceeds Four Million for First Time." UNHCR, July 9, 2015. https://www.unhcr.org/news/press/2015/7/559d67d46/unhcr-total-number-syrian-refugees-exceeds-four-million-first-time.html.

Vaughan-Williams, Nick. *Border Politics: The Limits of Sovereign Power*. Edinburgh: Edinburgh University Press, 2009.

Vergara-Camus, Leandro. "The MST and the EZLN Struggle for Land: New Forms of Peasant Rebellions." *Journal of Agrarian Change* 9, no. 3 (2009): 365–91.

Virno, Paulo. "Virtuosity and Revolution: The Political Theory of Exodus." In *Radical Thought in Italy: A Potential Politics*, edited by Paolo Virno and Michael Hardt, 189–210. Minneapolis: University of Minnesota Press, 1996.

Walia, Harsha. *Undoing Border Imperialism*. Oakland: AK Press, 2013.

Wang, Jackie. *Carceral Capitalism*. South Pasadena, CA: Semiotext(e), 2018.

Warren, Scott. "In Defense of Wilderness: Policing Public Borderlands." *South Atlantic Quarterly* 116, no. 4 (2017): 863–72.

Watts, Jonathan. "New Round of Oil Drilling Goes Deeper into Ecuador's Yasuní National Park." *The Guardian*, January 10, 2018, sec. Environment. https://www.theguardian.com/environment/2018/jan/10/new-round-of-oil-drilling-goes-deeper-into-ecuadors-yasuni-national-park.

Watts, Vanessa. "Indigenous Place-Thought and Agency Amongst Humans and Non Humans (First Woman and Sky Woman Go On a European World Tour!)." *Decolonization: Indigeneity, Education & Society* 2, no. 1 (April 5, 2013). http://decolonization.org/index.php/des/article/view/19145.

Wedekind, Jonah. "Jason W. Moore: Political Ecology or World-Ecology?," *Undisciplined Environments* (blog), January 12, 2016. https://undisciplinedenvironments.org/2016/01/12/jw-moore-politicalecology-or-worldecology/.

Weizman, Eyal. *The Least of All Possible Evils: Humanitarian Violence from Arendt to Gaza*. London and New York: Verso Books, 2011.

Whyte, Kyle Powys. "Our Ancestors' Dystopia Now: Indigenous Conservation and the Anthropocene." In *Routledge Companion to the Environmental Humanities*, edited by Ursula K. Heise, Jon Christensen, and Michelle Niemann, 206–15. New York: Routledge, 2016.

Wildcat, Daniel R. *Red Alert!: Saving the Planet with Indigenous Knowledge*. Colorado: Fulcrum Publishing, 2010.

Williams, Alex, and Nick Srnicek. "#ACCELERATE MANIFESTO for an Accelerationist Politics." *Critical Legal Thinking* (blog), May 14, 2013. http://criticallegalthinking.com/2013/05/14/accelerate-manifesto-for-an-accelerationist-politics/.

Williams, Eric. *Capitalism and Slavery*. Chapel Hill, NC: University of North Carolina Press, 2014.

Wintour, Patrick. "'Go Home' Billboard Vans Not a Success, Says Theresa May." *The Guardian*, October 22, 2013, sec. Politics. https://www.theguardian.com/politics/2013/oct/22/go-home-billboards-pulled.

Wise, Justin. "Women Sentenced to Probation after Leaving Food and Water for Migrants in Arizona Desert." *The Hill*, March 3, 2019. https://thehill.com/latino/432382-women-sentenced-to-probation-after-leaving-food-and-water-for-migrants-in-arizona.

Woods, Clyde. *Development Drowned and Reborn: The Blues and Bourbon Restorations in Post-Katrina New Orleans*. Edited by Laura Pulido and Jordan T. Camp. Athens, GA: University of Georgia Press, 2017.

Wynter, Sylvia, and Katherine McKittrick. "Unparalleled Catastrophe for Our Species? Or, to Give Humanness a Different Future: Conversations." In *Sylvia Wynter: On Being Human as Praxis*, edited by Katherine McKittrick, 9–89. Durham: Duke University Press, 2015.

Zeitoun, Mark, and Jeroen Warner. "Hydro-Hegemony—a Framework for Analysis of Trans-Boundary Water Conflicts." *Water Policy* 8, no. 5 (2006): 435–60.

Zhou, Cissy. "Man vs Machine: Automation Starting to Take a Toll on China's Workforce." *South China Morning Post*, February 14, 2019. https://www.scmp.com/economy/china-economy/article/2185993/man-vs-machine-chinas-workforce-starting-feel-strain-threat.

INDEX

See also: Settler colonialism

D

E

F

See also: Healthcare

Nineteen Eighty Four (George Orwell), 32

Nixon, Rob, 196

No Future (Lee Edelman), 145–46, 150

No More Deaths/No Más Muertes, 25, 68–69

No One Is Illegal, 51

O

O'Connor, James, 119–24, 134

Obama, Barack, 199

Occupy Sandy, 238

Occupy Wall Street, 210, 212

Octavia's Brood (adrienne maree brown and Walidah Imarisha), 31–32

Oikeios, 135–39

Oikonomia, 57–60, 66, 140, 193

Oil, 70, 136, 189, 198–99, 218, 220

Operation Triton, 50

Optimism, 12, 31, 74, 230

Organic agriculture, 42, 105, 111–16, 179

Ostrom, Elinor, 80–81

Overproduction, 121–22, 128
See also: underconsumption, underproduction.

P

Palestine, 153–54, 213, 215

Paris Agreement, 1, 8, 189, 219n7

Paris Conference (United Nations Climate Change Conference), 49, 219

Peasants, 76, 101, 106–08, 129

People, the, 48, 60–61, 90, 97, 192, 200, 207–15
See also: *Pueblo*

People's Climate March, 209–15

Pessimism, 31, 175

Piketty, Thomas, 222, 225

Place, 7, 24, 43, 74–76, 89–91, 95–96, 135, 147, 179, 182, 186

Polanyi, Karl, 103–04

Police, 11, 20, 23, 25, 29, 35, 40, 52, 64, 190, 200, 233, 235

Pollution, 2, 9, 19, 44, 51, 109, 112, 122, 124, 136, 138, 141, 175, 232

Poor Laws, 84, 100

Population, 8–9, 11, 13, 31, 52, 75, 79–85, 88n3, 100, 102, 146, 167–171, 174–75, "surplus population" (Marx), 8–9, 66, 183

See also: Malthus, Thomas, Malthusianism

Populism, 38–39, 48, 56, 57, 60–62, 88, 90, 197, 207–15
See also: counterpopulism

Poverty and Famines (Amartya Sen), 101–02

Profit, 23, 103, 121–22, 127, 137, 182, 196–98,

Project Cybersyn, 181

Prometheanism, 73–74, 76–77, 126, 128–29, 203

Puar, Jasbir K., 58, 93, 156

Pueblo, 212, 237

Puerto Rico, 232, 235, 237
See also: Hurricane Maria

Q

Queerness, 58, 69, 96n31, 139, 146–47, 150–52, 156, 161–62, 164, 202

R

Race, 10, 32, 56–59, 66, 70, 153

Racial capitalism, 10, 28, 160, 186, 192

Racial violence, 10–11, 13–14, 25, 51, 57–58, 64, 91, 233

Rangasami, Amarita, 101

Ratzel, Friedrich, 89

Reinsurance, 221

Refuge, 24, 69

Refugees, 11, 20–22, 50n11, 51, 56, 96, 155
See also: Borders
See also: Climate Refugees
See also: Migrants, Migration
See also: No Borders

Renewable energy, 7, 70, 218, 221n15

Reproductive futurism, 58, 145–147, 149–57, 202
See also: Child, the
See also: Edelman, Lee

Romanticism, 6, 7, 14, 42, 73–74, 77, 92, 140, 178–79, 202, 224

Roosevelt, Theodore, 93–94

Ruins, 3, 5–6, 9–10, 15, 30, 131, 157, 240

S

Sahelian famine (1973), 101

Sanctuary Cities, 68

San Francisco earthquake (1906), 4, 233, 235

Sand Creek Massacre (1864), 154

Schumacher, E.F., 130

ABOUT OUT OF THE WOODS COLLECTIVE

Out of the Woods is a transnational political research and theory collective, a loose grouping of decolonial, small-c communist, antiracist queer-feminist thinkers working together to think through the problem of ecological crisis.

Out of the Woods began in 2014 as a collective investigation into the various historical, contemporary, and future relationships between capitalism and climate change. We are exhausted by the way in which hegemonic ecological politics in the so-called Global North oscillates between a return to a romantic naturalism with reactionary tendencies and a pragmatic green capitalism. We are also inspired by and seek to amplify real movements abolishing the present state of things: survivors of Hurricane Katrina, Indigenous water protectors opposing transnational pipelines, migrants fighting border imperialism, struggles for Black lives in Ferguson and safe water in Flint.

We recognize the need for a much wider lens on both the breadth of ecological crisis and the historical forms through which it perpetuates exploitation, dispossession, exhaustion, and maybe insurrection. Out of the Woods is evolving with the intention of further multiplying our positions against homogeneity. We invite you to contact us, and to think, write, and struggle with us. As disaster engulfs spaces and times around us, what will it take to get out of the woods, together?

Out of the Woods Collective
outofthewoodscollective@gmail.com

ABOUT COMMON NOTIONS

Common Notions is a publishing house and programming platform that advances new formulations of liberation and living autonomy.

Our books provide timely reflections, clear critiques, and inspiring strategies that amplify movements for social justice.

By any media necessary, we seek to nourish the imagination and generalize common notions about the creation of other worlds beyond state and capital. Our publications trace a constellation of critical and visionary meditations on the organization of freedom. Inspired by various traditions of autonomism and liberation—in the U.S. and internationally, historically and emerging from contemporary movements—our publications provide resources for a collective reading of struggles past, present, and to come.

MORE FROM COMMON NOTIONS

Towards the City of Thresholds
Stavros Stavrides
978-1-942173-09-0
$20.00
272 pages

In the Name of the People
Liaisons
978-1-942173-07-6
$18.00
208 pages

*For Health Autonomy: Horizons
of Care Beyond Autonomy—
Reflections From Greece*
CareNotes Collective
978-1-942173-14-4
$15.00
144 Pages

Abolishing Carceral Society
(*Abolition: A Journal of Insurgent
Politics*)
Abolition Collective
978-1-942173-08-3
$20.00
256 pages